P9-CJP-671

WITHDRAWN

MY EINSTEIN

MY EINSTEIN

*Essays by Twenty-four of the
World's Leading Thinkers on
the Man, His Work, and
His Legacy*

EDITED BY JOHN BROCKMAN

CONTRA COSTA COUNTY LIBRARY

Pantheon Books New York

3 1901 04017 4197

Copyright © 2006 by John Brockman

All rights reserved. Published in the United States
by Pantheon Books, a division of Random House, Inc.,
New York, and in Canada by Random House
of Canada Limited, Toronto.

Pantheon Books and colophon are registered trademarks of
Random House, Inc.

Library of Congress Cataloging-in-Publication Data
My Einstein : essays by twenty-four of the world's leading
thinkers on the man, his work, and his legacy / edited by
John Brockman.

p. cm.

ISBN 0-375-42345-1

1. Einstein, Albert, 1879–1955—Anniversaries, etc.
2. Physicists—Germany—Biography. 3. Physicists—United
States—Biography. 4. Jewish scientists—Germany—
Biography. 5. Jewish scientists—United States—Biography.
I. Brockman, John [date]

QC16.E5A5 2006

530'.092—dc22

[B] 2005048286

www.pantheonbooks.com

Printed in the United States of America

First Edition

2 4 6 8 9 7 5 3 1

To Sidney Coleman, a true seeker

Contents

CONTENTS

CONTENTS

Introduction

JOHN BROCKMAN

Most readers of this volume will already know quite a bit about Albert Einstein, whose centennial we celebrated in 2005—the year not of his birth but of his annus mirabilis, when he produced five papers that have forever altered our perception of reality.

But to reprise the basic facts: Einstein was born on March 14, 1879, in Ulm, Württemberg, Germany, and died on April 18, 1955, in Princeton, New Jersey. The five 1905 papers are his University of Zurich doctoral dissertation on the determination of molecular dimensions and the four more famous ones, listed here in order of their submission to *Annalen der Physik:*

- on light quanta and the photoelectric effect ("On a Heuristic Point of View About the Creation and Conversion of Light"—this is the work for which he was awarded the Nobel Prize in 1921);

- on Brownian motion ("On the Movement of Small Particles Suspended in a Stationary Liquid Demanded by the Molecular-Kinetic Theory of Heat");

- and two papers on special relativity ("On the Electrodynamics of Moving Bodies" and "Does the Inertia of a Body Depend on Its Energy Content?," in which appears his famous equation of matter and energy, $E = mc^2$).

In the years following this spectacular production, Einstein devoted himself chiefly to incorporating the gravitational force into his theory of relativity, and in 1916 published "The Foundations of General Relativity Theory." In that paper and one the following year entitled "Cosmological Considerations of the General Theory of Relativity," he, in a sense, took on the universe itself. It is in the latter paper that he introduces the cosmological constant, later repudiated by him as his "greatest blunder" but now very much back in favor with some cosmologists as a means of describing the recently discovered acceleration of the universal expansion.

Einstein was clearly the most important person of the twentieth century. He achieved an iconic status that (some would say unfortunately) transcends even the heights of his scientific genius. We all feel that we know him. We all think about him in different ways.

I have therefore asked the contributors to *My Einstein* to address the following questions: Who was Einstein to you? What difference did he make to your worldview, your ideas, your science? How did Einstein influence you personally? Who is your Einstein?

The two dozen essayists in *My Einstein* are among the

world's leading theoretical and experimental physicists, science historians, and science writers. But this is not just a book about physics. It is a collection of personal narratives, providing a unique window into how these thinkers assess Einstein's scientific and philosophical legacy and his particular influence on their own lives and work. They are:

ROGER HIGHFIELD on the Einstein myth;
JOHN ARCHIBALD WHEELER (the only one who actually knew Einstein, though the Nobel laureate Leon Lederman once met him briefly) on their meetings in Princeton, Wheeler on the Princeton physics faculty and Einstein at the Institute for Advanced Study;
GINO C. SEGRÈ, LEE SMOLIN, and ANTON ZEILINGER on Einstein's difficulties with quantum theory;
GEORGE F. SMOOT and PETER GALISON on Einstein's blending of pure thought and physical observation;
LEON M. LEDERMAN on the special theory of relativity;
CHARLES SEIFE on Einstein's use of the gedanken-experiment;
FRANK J. TIPLER on why Einstein should be seen as a scientific reactionary rather than a scientific revolutionary;
GEORGE DYSON on growing up in Princeton and his friendship with Helen Dukas, Einstein's longtime amanuensis;

COREY S. POWELL on the philosophical underpin-
nings of Einstein's use of the word "God";

STEVEN STROGATZ, GEORGE JOHNSON, and JEREMY
BERNSTEIN on how Einstein turned them on to
physics in their early years;

LEONARD SUSSKIND on Einstein's way of thinking;

JANNA LEVIN and MARIA SPIROPULU on how he is
perceived among physicists in academe today;

MARCELO GLEISER on Einstein's new world of
mysterious properties and bizarre effects;

PAUL C. W. DAVIES, LAWRENCE M. KRAUSS, and
ROCKY KOLB on the accelerated expansion of the
universe and the revival of Einstein's cosmological
constant;

RICHARD A. MULLER on the mysterious nature of
time;

PAUL J. STEINHARDT on a new cosmology involving
a cyclic universe and its relation to Einstein's
cosmological thought.

And me? Who is my Einstein?

I remember the moment I found out about Einstein's
death, brought up short by a headline at a kiosk in an
underground station of Boston's MTA. I was fourteen at
the time. It was a shattering moment, in which I felt genuine grief and loss.

By then my family had moved to the relative peace and
quiet of the suburbs, but the first ten years of my life had
been marked by learning survival tactics in the "other"

Boston—miles away from the graceful sailboats on the Charles River, the gleaming golden dome of the State House on Beacon Hill, the serene beauty of Harvard, the bold architecture of MIT.

I grew up in Dorchester in the 1940s. It was a tough, gritty neighborhood, where, before World War II, Father Charles E. Coughlin, the infamous "Radio Priest," had regularly sent sound trucks up and down the streets spreading his anti-Semitic gospel. This assault had helped to turn Dorchester into a battleground between the Irish kids and the greatly outnumbered Jewish kids. Our three-block walk to the William E. Endicott School on Blue Hill Avenue was a daily obstacle course—my brother Philip, three years my senior, had to defend himself while also protecting me. Our sense of perilous vulnerability was heightened by the realization that anyone with any kind of civic authority—be it a teacher, trolley-car conductor, or cop—seemed always to have a name like Flaherty, O'Reilly, or McCormack.

The fights we got into were almost always part of a broader history lesson: Philip and I discovered that we were personally responsible for the death of the Second Person of the Holy Trinity. We tried reasoning. None of our arguments—Jesus was a rabbi, who prayed in Hebrew and preached in a synagogue; his mother looked like our mother, not like their mothers—seemed to impress these furious young toughs.

But we did have a secret weapon, the most powerful kind, one we realized they would never possess or even understand. On more than one occasion when we limped home from battle, while tending to our bloody noses, cuts, and scrapes, our mother would buck us up, vigorously fighting bigotry in kind:

"Look at them! What the hell do they have? They bake a ham on Sunday and eat it all week! The men don't bathe! The women leave their babies in carriages outside the bars! But look what we have!" Her blue eyes beamed strength, certainty, and pride as she dabbed at our bruises. "What we have, they will never have. We have . . . Einstein!"

My mother was right. We had Einstein with us, as we made our way up through the terrifying school system and investigated what the public library had to offer. He gave us permission to think big thoughts, to explore intellectually the remotest corners of existence. He allowed us to appreciate, to embrace, the life of the mind. He was always with us. We did have Einstein; we still have Einstein.

My brother Philip became a research physicist and recently retired after a long career at NASA. He is now Distinguished Research Associate at NASA and a recipient of its Exceptional Service Medal. As for me, today I am fortunate to work with, and count among my friends, leading cosmologists, particle physicists, and string theorists, all of them to some degree Albert Einstein's heirs. You could say that I'm very lucky . . . but maybe luck had nothing to do with it. You see, I had Einstein—my Einstein.

MY EINSTEIN

Einstein When He's at Home

ROGER HIGHFIELD

ROGER HIGHFIELD is the science editor of the *Daily Telegraph* in London. He has carried out research at Oxford University and the Institut Laue-Langevin in Grenoble, where he became the first to bounce a neutron off a soap bubble. He is the author of *Can Reindeer Fly?: The Science of Christmas; The Science of Harry Potter: How Magic Really Works;* and coauthor (with Paul Carter) of *The Private Lives of Albert Einstein* and (with Peter Coveney) of *Frontiers of Complexity* and *The Arrow of Time,* a best seller that has been translated into more than a dozen languages.

Here is the canonical Einstein: He begins life as a dullard and a dyslexic, yet he overcomes these obstacles to help lay the foundations of quantum theory, to change our view of space, and to transform time. Despite his towering achievements, he shows great humility. He pokes his tongue out for the cameras. He is disheveled. He hates socks. He is an eccentric genius with a warm heart. He is a pacifist (except when it comes to the Nazis). His face is wise and lined, his hair is white and wild; some call it a mane or even a halo. When describing the universe, Einstein resorts to religious terms. He has the aura of a saint. But he also has a dark secret: he invented the atomic bomb.

The popular image of Einstein as archetypal eccentric boffin dates to half a century after the first flowering of his astonishing creative genius. The tangle-haired sage whose image has graced thousands of posters, coffee mugs, and T-shirts is an Einstein well past his scientific best, a faded version of the original. We should bury the sockless dust-ball who rolled around Princeton and restore the creative Einstein.

This is the young Einstein, whom Paul Carter and I attempted to portray in our 1993 book *The Private Lives of Albert Einstein*, after conversations with relatives and with scholars such as Jürgen Renn, John Stachel, and Robert Schulmann. This is the passionate Einstein. This Einstein had a muscular and powerful build despite his indifference to most forms of exercise. He had regular features, warm brown eyes, a mass of curly black hair, and a raffish mustache. He was good-looking and enjoyed the company of women. They enjoyed his company, too. And, of course, he was a genius. That much was obvious early.

Einstein was not stupid as a child. He did repeat himself, but he was not dyslexic, as is often asserted. Classmates at his primary school taunted him with the nickname "Biedermeier" ("Honest John"), most likely because of his blunt manner. But his mother, Pauline, wrote in August 1886 that the seven-year-old was at the top of his class "once again" and had received a "splendid" report card. He was brought up in a family that made its living from electrical engineering, an advanced technology of the day.

Despite his love of a religious turn of phrase, Einstein found it impossible to conceive of a personal deity and had no belief in an afterlife. He has said that his reading of popular science ended his "religiosity" abruptly, at the age of twelve. He decided that the stories of the Bible could not be true and became a fanatical freethinker, convinced he had been fed lies.

He did not invent the atom bomb. He did transform our view of space and time. His great scientific works began with a creative outpouring in 1905, when he was just twenty-six years old. Like almost every other scientist and mathematician, he was at his most productive in his early years.

No one knew the real Einstein better than his first wife, Mileva Marić. Their marriage, from 1903 to 1919, spanned the most important years of his life, yet Mileva is a shadowy figure in many Einstein biographies. Because of a lack of letters from that period and their uneasiness about his first marriage and its many failings, the traditional biographers tended to focus on Einstein's later years. In these hagiographies, in which the assumption is made that a great scientist must have an unwrinkled private life, the old Einstein prevails.

A chance to see Einstein afresh came when his son Hans Albert died in July 1973. In a shoebox in a drawer at his home in Berkeley, California, was family correspondence, including love letters from Einstein to Mileva. The collection was so sensitive that the executors of Einstein's estate had gone to court to stop Hans Albert from publishing it; they argued that not even Einstein's son, to whom many of the letters were addressed, should be allowed to reveal such intimate material. Only in recent years have the letters been published, and only now can we see Einstein in his prime, warts and all.

The young Einstein would moan to Mileva that his mother and sister were crass, petty, and philistine. He complained about the "mindless prattle" of his mother's friends and relatives. His Aunt Julie was a "veritable monster of arrogance." His relatives and their "hangers-on" were "people gone soft," turned "moldy," whose lives were empty and whose minds had atrophied.

The young Einstein was no respecter of scientific reputations, either—not least because he had been shunned by the establishment after he graduated from the Swiss Federal Polytechnic School (now called the Eidgenössische Technische Hochschule, or ETH) in 1900. The work of Paul Drude, one of the leading theorists of the day, was "stimulating and informative" but lacked clarity and precision. Einstein sent Drude a series of objections to his electron theory of metals (in which various properties are explained in terms of an electron gas). Having come up with a similar theory, he felt it quite proper to approach Drude as an equal and point out his "mistakes." He threatened to "make it hot" for Drude by publishing an attack on Drude's theory. "Unthinking respect for authority is the greatest enemy of truth," he declared.

Einstein's comments on his instructors at the Polytechnic School were equally biting. One taught clearly but too superficially; another was brilliant and profound but an impenetrable pedant. When Einstein struggled to find a job, he accused his old physics lecturer of thwarting his career by spreading bad opinions of him. He went out of his way to annoy the head of a boarding school where he worked for a short time, and he told Mileva, "Long live impudence. It's my guardian angel in this world."

These vivid letters do much to undermine the image of the genial sage. He could be sweet, funny, and charming, but he could be acerbic, too. Einstein exuded charisma and a relaxed charm. He was a flirt. Far from the shaggy-haired figure of later years, he possessed what a friend of Elsa, his second wife, would describe as "masculine good looks of the type that played havoc at the turn of the century."

He met Mileva in 1896, when she switched to Section VI-A of the Polytechnic School, reading, like Einstein, for a diploma that would qualify her to teach mathematics and physics at secondary schools. Mileva, almost twenty-one, was three and a half years older than Einstein and the only woman to join Section VI-A that year (and only the fifth altogether). A romance developed (not Einstein's first). This is not a love story about Albert and Mileva but about "Johonesl" ("Johnnie") and "Little Doll" ("Dollie," as well as his tiny witch, his itty-bitty frog, his dear kitten, his little street urchin, his dear little angel, his little right hand, his dearest little child, his tiny black girl). On August 20, 1900, Johnnie wrote his Dollie a daft and endearing dialect poem, which includes the verse:

> Oh my! That Johnnie boy!
> So crazy with desire,
> While thinking of his Dollie,
> His pillow catches fire.

The popular image of an elderly Einstein does not even suggest the possibility of a fling, let alone this engaging bit of lovestruck silliness. Two years later, the relationship became serious. Johnnie and Dollie conceived a love child during a trip to the Splügen Pass, near Como. Mileva gave birth sometime around the end of January 1902, yet there is no evidence that Einstein and Lieserl, his daughter, ever set eyes on each other. Einstein was never to talk of Lieserl publicly. She might have been erased from history had it not been for the discovery of the cache of love letters. Her fate is not known for certain. Perhaps Lieserl's birth posed a threat to Einstein's new start as a patent examiner in Bern. He had gained Swiss citizenship only a year earlier, and the stigma of an illegitimate child would have harmed his prospects. She was probably surrendered for adoption. Though understandable, this is hardly behavior one might have ascribed to the latter-day saint.

In Bern, Einstein engaged in high jinks and schoolboy pranks. He offered private tutoring in mathematics and physics, and among his pupils was Maurice Solovine, an ebullient Romanian studying at Bern University. Later the two were joined by Conrad Habicht, the uptight scion of a bank director. They constituted themselves, with mock formality, as the Olympia Academy and discussed philosophical issues, Einstein taking the lead. The door of his and Mileva's apartment was adorned with a tin plaque reading "Albert Ritter von Steissbein, President of the Olympia Academy." "Ritter von Steissbein" might loosely be translated as "Knight of the Backside." On one occasion, Solovine skipped a meeting at his own lodgings, and Einstein and Habicht took revenge by smoking furiously (Solovine hated tobacco) and piling all his belongings, from furniture to crockery, on his bed.

Each paper that Einstein produced in his annus mirabilis is the final consequence of a long chain of work by masters of classical physics—Ludwig Boltzmann, Max Planck, Hendrik Lorentz. However, Einstein had sufficient distance from their way of thinking to interpret their research from a new perspective—with revolutionary results. Far from earning him instant acclaim, the papers were at first largely ignored. According to his sister, Maja, Einstein had expected immediate criticism of his relativity theory; instead there was silence, and he was disappointed. The exception was the highly influential Planck, who began lecturing on the theory, firing the imagination of his assistant Max von Laue, one of the first scientists to pay a call on the unknown author in Bern. Von Laue was confronted not with a sage but with a garrulous young man. He found Einstein's appearance so unprepossessing when he first arrived at the patent office that he let the young man walk past him ("I could not believe he could be the father of the relativity theory"). He was equally unimpressed by the cheap cigar Einstein gave him, and as they crossed a bridge over the Aare he surreptitiously threw it into the river.

Einstein's private life was not nearly as successful as his science, though one would not gather this from his early biographers. He could be harsh. When, in late 1932, his son Eduard was admitted for the first of many stays at the Burgholzli mental institution in Zurich to be treated for schizophrenia, Einstein is said to have remarked, "Who knows if it would not have been better if he had left the world before he had really known this life." Einstein also had a streak of misogyny. Of a woman who he felt was tormenting "a great artist" of his acquaintance he declared: "You know, that is a creature I could kill in cold blood. I'd like to put a rope around her neck and tighten it until her tongue lolled out." This dramatic statement was accompanied by the appropriate gestures.

Einstein had misgivings about matrimony. It must have been invented "by an unimaginative pig" and was "slavery in a cultural garment." He argued from firsthand experience that marriage was incompatible with human nature, claiming that 95 percent of all men, and probably as many women, were not monogamous by nature. He once joked that he preferred "silent vice to ostentatious virtue." Marriage reduced free human beings to mere articles of property and was "the unsuccessful attempt to make something lasting out of an incident." Asked on one occasion whether it was permissible for Jews to marry non-Jews, he replied with a laugh, "It's dangerous—but then all marriages are dangerous."

He told his lover—his cousin Elsa—that Mileva was "an unfriendly, humorless creature who herself has nothing from life and who undermines others' joy of living through her mere presence." She was "the sourest sourpot there has ever been," a plagued individual who gave their home the atmosphere of a cemetery. Her jealousy was a pathological flaw typical in a woman of such "uncommon ugliness." Then again, Mileva had good reason to be unhappy. When in 1916 Einstein demanded a divorce, she suffered a physical and mental breakdown. The divorce was finally issued in February of 1919, and Einstein and Elsa were married the following June.

Within a few months, he had become celebrated across the planet: "REVOLUTION IN SCIENCE / NEW THEORY OF THE UNIVERSE / NEWTONIAN IDEAS OVERTHROWN" thundered the *Times* of London on November 7, 1919. "LIGHTS ALL ASKEW IN THE HEAVENS / MEN OF SCIENCE MORE-OR-LESS AGOG / EINSTEIN THEORY TRIUMPHS," announced the *New York Times* two days later. The accompanying reports revealed the findings of two British expeditions to observe a solar eclipse. Scientists in northern Brazil and on the island of Príncipe off the west coast of Africa had witnessed the bending of starlight predicted by his general theory of relativity. The results caused a sensation at the Royal Society, whose president hailed relativity as perhaps the most momentous product of human thought. Biographer Abraham Pais has called this "the birth of the Einstein legend."

The architect of Einstein's summer house in Caputh, near Berlin, where the now world-famous professor spent much time from 1929 through 1932, noted that women were drawn to him like iron filings to a magnet and that Einstein responded eagerly. Various liaisons developed, some of them casual, a few intimate, all wounding to Elsa, whom he provoked into the same jealous furies that he had complained of in Mileva.

Reporters scrambled to interview the man behind the theory and were enchanted to find a wild-haired eccentric of rumpled charm and displaying a mocking sense of humor. He became a media sage, courted the world over. During a trip to Geneva he was mobbed by young girls, one of whom tried to snip off a lock of his hair. Babies were named after him, as were a telescope and a brand of cigars, and a torrent of letters began to arrive. They continued for the rest of his life: letters from well-wishers, religious nuts, spongers begging for money, pressure groups seeking endorsements, children wanting help with their homework—even one from a little girl asking, "Do you exist?"

The young Einstein who had achieved so much and whose efforts climaxed with his general theory in 1915 no longer did exist, of course. As Einstein entered his forties, he turned to the quest for a unified field theory—a set of equations that would marry the laws of gravity and electromagnetism. These were then believed to be the two fundamental forces in nature, so a theory explaining both would solve all nature's puzzles. As his fiftieth birthday approached in 1929, reports circulated that he was on the verge of a great discovery, and when his latest paper emerged it was printed in full by the *New York Herald Tribune*. In London it was displayed in the window of Selfridges department store and drew large crowds. All this was a tribute to the power of Einstein's name, but in fact the paper's thirty-three equations were meaningless to the layman and solved only a few preliminary problems of fiendish technical complexity.

At the Institute for Advanced Study in Princeton, Einstein pursued unification with his assistant Walther Mayer and then with Valentin Bargmann and Peter Bergmann.

They were working together in December 1936 in his house at 112 Mercer Street as Elsa was dying in the next room. Her anguished cries unnerved Bergmann, but Einstein was absorbed. When Einstein seemed to be on the right track, he would declare, "This is so simple that God could not have passed it up." In 1949 he outlined his ideas on unification for an appendix to the third edition of his book *The Meaning of Relativity;* this publication became a worldwide event, mainly because of its coincidence with his seventieth birthday.

After Elsa's death, Einstein's love life became less complicated. He gave up the affairs and pursued serial monogamy, first with Margarita Konenkova, a Russian spy fifteen years his junior (he was ignorant of her occupation). His last girlfriend was Johanna Fantova, former curator of maps in Princeton's Firestone Library, whom he had first met in 1929. She was twenty-two years his junior; he told her he was still a revolutionary and "a fire-spewing Vesuvius."

On his own deathbed, Einstein called for his most recent calculations on the unification of gravity and electromagnetism. Yet his final scientific mission was fundamentally misconceived, according to (among many others) the physicist who later came to live in his house on Mercer Street, the 2004 Nobelist Frank Wilczek. Einstein had ignored emerging evidence that gravity and electromagnetism were not nature's only fundamental forces. Today those who venerate Einstein's real achievements are baffled—as much by his errant direction in his twilight years as by the saintly Einstein who lives on in legend.

The Freest Man

GINO C. SEGRÈ

GINO C. SEGRÈ, a noted theoretical physicist from a distinguished family of physicists, was born in Florence and raised there and in New York City. He has held visiting professorships at both MIT and Oxford, and is currently a professor in the Department of Physics and Astronomy at the University of Pennsylvania. From 1987 to 1992 he was the chair of the department, and in 1995 he became the director of theoretical physics for the National Science Foundation. He is also the author of *A Matter of Degrees: What Temperature Reveals About the Past and Future of Our Species, Planet, and Universe.*

"He was the freest man I have ever known" is how Abraham Pais characterizes Albert Einstein in his biography. Pais, a distinguished physicist and a colleague and friend of Einstein's at the Institute for Advanced Study in Princeton, knew him well—or as well as a young man could know the great figure in his declining years. They frequently strolled together—Pais would accompany Einstein back from the latter's office at the Institute to his house on Mercer Street—discussing quantum theory and relativity as they took their exercise. Pais was also often the third person in the room as Niels Bohr, visiting Princeton, continued his twenty-year debate with Einstein about the probabilistic interpretation of quantum mechanics, the two of them in a haze of smoke as they puffed away on their pipes. Later, Pais elaborated on that initial statement, saying that by "freest man" he meant that Einstein was more "master of his own destiny" than anyone Pais had ever known.

This freedom, manifested as independence, sets Einstein apart from other major twentieth-century figures. The refrain Janis Joplin immortalized—"Freedom's just another word for nothing left to lose"—does not apply here. In Einstein's case, there was a great deal to lose. His personal, political, and intellectual freedom came at a price, but one he was willing to pay because the cost of not having it was unacceptable.

Einstein's predilection for independence was evident early. In his teens, the family's failing business caused them to relocate from Munich to northern Italy; young Einstein remained behind in order to finish his studies and visited them when he could. However, at age sixteen, tired of Germany and the rigidities of its school system, he left the Munich *Gymnasium*, renounced his German citizenship, and continued his schooling at the Eidgenössische Technische Hochschule in Zurich. Five years later, in 1901, he obtained Swiss citizenship. Though Einstein continued to regard himself as Swiss and traveled with a Swiss passport, he acquired de facto German (or more correctly, Prussian) citizenship when he became a member of the Prussian Academy of Sciences in 1913, a position he accepted on condition that he be allowed to remain a Swiss citizen as well. This dual nationality would last until 1923, when he was required to formally accept German citizenship.

In 1914, at the start of World War I, Einstein was the director of the Kaiser Wilhelm Institute of Physics and a professor at the University of Berlin. Two months after the war began, his Berlin colleague Max Planck joined thousands of other German academics in an appeal entitled "To the Cultured Peoples of the World" (*An Die Kulturwelt! Ein Aufruf*), a document denying charges that German troops had committed atrocities in Belgium, destroyed works of art, burned the library in Louvain, and shot civilians. It ends by asserting that, contrary to accusations, the land that had given birth to Beethoven and Goethe knew how to respect Europe's cultural heritage. Shortly thereafter, G. F. Nicolai, another Berlin professor, drafted a counter-manifesto deploring the previous statement's militaristic and nationalistic tone. His rebuttal exhorted Europeans to unite in order to avoid the horrors that were sure to follow. Warning of the dangers of warfare, his plea includes this prophetic note: "It is therefore not only desirable, but a dire necessity that educated men of all nations make their influence felt, so that whatever the outcome of the war, the terms of the peace do not become the source of future wars." This counterappeal was not published, because it managed to garner only four signatures. One of them was Einstein's.

In the meantime, he was working harder than ever on his research. During the war, he would produce the general theory of relativity, regarded by many as the single most perfect theory ever created by one individual and a monument that will endure for centuries. Fond as he was of his Switzerland, Einstein had no illusions about how attached his adopted country was to him. As he quipped, "If relativity is correct, the Swiss will say I am Swiss and the Germans will say I am German. If it is wrong, the Swiss will say I am German and the Germans will say I am a Jew." The ambivalence of his national status led occasionally to quasi-comic diplomatic situations: Since he was unable to be in Sweden in 1922 to receive the Nobel Prize, the award was made in Einstein's name to the German ambassador to Sweden but handed over to him in Berlin by the Swiss ambassador, both countries now obviously eager to claim him as one of their own.

But though his national loyalties may have been in question, his identification as a Jew was clear. In the Germany of the 1920s, while his brethren were being threatened by anti-Semitism, he was outspoken against the growing excesses, increasingly so as the country veered in a direction he deplored. Other Jews thought that the atmosphere of repression would lessen and the increasing power of Nazism would pass; Einstein was unconvinced. He became an influential supporter of Zionism, helping to raise funds for Jewish causes. In 1952, three years before he died, he was even offered the presidency of the new nation of Israel—a figurehead position but a deeply symbolic one. Moved, he nevertheless declined, replying, "I know a little about nature and hardly anything about men."

And yet Einstein had renounced Judaism as an official identity when he was sixteen years old. His God was a more abstract deity, one of reason and order in the world rather than a figure to be prayed to. Occasionally he would invoke a supreme being when talking about physics, mockingly hinting at a personal relationship—as in his famous phrase "God does not play dice," a rejection of the probabilistic interpretation of quantum mechanics as the ultimate description of the microscopic world. In their ongoing debate, Bohr—when he heard that phrase, and he often did—would reply to Einstein in a similar vein: "Nor is it our business to prescribe to God how He should run the world." At other times Einstein would simply refer, half-jokingly, to *Der Alte*, the Old One.

Though he valued advice from the Old One, Einstein placed a higher value on the personal freedom that allowed him to step back in ways others found hard, to shake off the ties to one or another country. In the 1920s and early 1930s, he traveled a great deal—to South America, to Japan, and three times to the United States for extended stays at the California Institute of Technology. Leaving in December of 1932 for one such trip, and sensing Germany's mood, he turned to his wife, telling her to have a good look at their Berlin house because she would never see it again. She regarded this statement as foolish, but on January 30, 1933, Hitler became Germany's chancellor. When the Einsteins returned to Europe in March of that year, their boat landed in Belgium, at Antwerp. There they went straight to the German Embassy in Brussels and renounced their German citizenship.

In an interview in a New York newspaper before returning to Europe, Einstein had said that "humanity is more important than national citizenship," a consistent theme in his life. Looking back at the great figures in physics who were contemporaries of Einstein, one cannot but be struck by how important the idea of a homeland was for them and how relatively unimportant it was for Einstein. Planck was German to the core, loyal even at times that were more than trying. He lost one son in World War I and another was executed by Hitler in the closing days of World War II. Yet he never viewed himself as anything but a true German. Bohr was a confirmed internationalist, but Denmark was his one and only home. Even Ernest Rutherford, born in New Zealand, kept his ties to that far-off colony he had left in his early twenties. Elevated to the peerage in 1931, he chose the title Lord Rutherford of Nelson, Nelson being the small secondary school he had attended as a youth. As he lay dying, he reminded his wife of the donation he wished to make to the school.

Several older Germans, those unable to start life over again successfully in another country, came back. Max Born, ousted by the Nazis in 1933, returned to Germany from Edinburgh when he retired in the 1950s, spending his last years near Göttingen, where he had taught twenty years earlier. Erwin Schrödinger traveled back to his native Austria for his final years. Others simply wanted to see old friends and familiar places. But for Einstein Europe was a closed chapter. Once he came to the United States, he never again set foot there, even for the briefest of visits, confiding to a friend, "I have never known a place that to me was a homeland [*Heimat*]. No country, no city has such a hold on me." Others clung to their idea of *Heimat* even when it rejected them; Einstein never did.

Most important, there is Einstein's intellectual freedom, his extraordinary ability to look at simple concepts and see them as nobody else ever had. To realize that the law of addition of velocities had to be modified, that there is a maximum velocity for the transmission of signals, that nobody before him had ever understood the concept of simultaneity—all this shows his willingness to think beyond boundaries. Einstein's special theory of relativity set him on the road to his greatest achievement, the general theory, which he labored on intermittently from 1907 until late in 1915. It all started with a simple idea, one that came to him as he sat at work in the patent office in Bern, already having achieved greatness but not yet recognized by a world whose mills ground exceedingly slowly.

Einstein's special theory of relativity had reinterpreted our measurements of space and time, but it was applicable only to observers moving relative to one another in the absence of all forces, so-called inertial frames of reference. It therefore could not be applied to the real world, where the force of gravity is pervasive, as Newton had explained. Now comes what Einstein later called his happiest thought: the interior of a sealed elevator in free fall toward the earth is an inertial frame. All objects on the inside act relative to one another as if the force of gravity were nonexistent (a sight by now familiar to those who have watched astronauts in space or trainees for such missions). Perhaps one should take this as a metaphor for the ultimate freedom: Einstein had discovered the means to do away with the force of gravity and yet take it into account. Boundaries regarded as inescapable could be removed after all.

The mass in Newton's $F = ma$, the relation between force and acceleration, is the same as the mass appearing in gravitational attraction. Einstein had now taken the first step along the bridge between his special theory and his general theory, the grand construct that would describe the action of forces as the bending of space and time. It took Einstein eight more years to fully implement his brilliant insight, to develop the necessary mathematical tools, and ultimately to write down a set of equations describing everything from the falling of an apple to the expansion of the universe, but it all started with that happiest thought.

By the 1920s, when the general theory of relativity was experimentally confirmed by the observed bending of light and he was awarded the Nobel Prize for his work on quantum theory, Einstein was probably the world's preeminent scientist, certainly its leading theoretical physicist and the inspiration for a rising generation. From 1927 on, however, he increasingly distanced himself from his fellow physicists by his unwillingness to accept what became their central belief—the Copenhagen interpretation of quantum mechanics. There was a great deal of sadness for them in this, particularly because Einstein had contributed as much as or more than any individual in bringing about the revolution spearheaded independently by Werner Heisenberg and Erwin Schrödinger. As Max Born said, "Many of us regard this as a tragedy—for him as he gropes his way in loneliness, and for us, who miss our leader and standard bearer." Yet my sense is that Einstein never minded this scientific isolation. Like the other ties he was willing to give up, the consensus in physics could also be dispensed with. Ultimately he stood apart from others, knew it, and invited this separation rather than seeking to overcome it.

This apartness is reflected in his work, lending it a strange quality of permanence, a weightiness that is absent in efforts by others. Pais comments that Einstein's papers "exude finality, even when they deal with a subject in flux. For example, no statement in the 1905 paper on light quanta needs to be revised in the light of later developments." I found this remark intriguing, because I wasn't then familiar with Einstein's writings. Unlike humanists or social scientists, natural scientists do not go back to original sources, any more than a medical doctor might, since the field moves rapidly and is propelled by new discoveries. We derive the essence of a finding from content, not style. I thought that Pais's remark might well apply to relativity theory, which Einstein had created in a sense out of whole cloth, but I doubted that it was true of his papers on quantum theory, a subject very much in transition as he was contributing to it. Curious, I went to a collection of historical papers and read his 1916 "On the Quantum Theory of Radiation." I knew roughly what the paper said, but like all physicists of my generation I had learned it from a textbook, not from the original. Approaching the article with some trepidation, I feared it would seem obscure to me, a disappointment. But there it was, clearer than any textbook, thrilling in an eerie way.

It is deeply moving to see Einstein, having just published his general theory of relativity, coming back to the quantum theory he had already done so much to advance.

Now he magisterially addresses emission and absorption of radiation and then goes on to incorporate in radiation the results of his 1905 special theory of relativity. In that theory, Einstein had shown that spatial and temporal intervals have no absolute meaning; one must think of them together. He also realized that energy and momentum are related to each other the way space and time are. In particular, since energy is conserved in collisions, he points out that momentum must be as well. A quantum of radiation transferring energy to a target will therefore also transfer momentum. At the end of his 1916 article, Einstein writes,

> [I]n general one restricts oneself to discussion of the energy exchange, without taking the momentum exchange into account. One feels justified in this. . . . [H]owever, such small effects should be considered on a completely equal footing with more conspicuous effects of a radiative energy transfer, since energy and momentum are linked in the closest possible way.

Of course the person who saw that linkage was Einstein.

Certain twentieth-century figures (Nelson Mandela comes to mind) have a curious way of seeming totally in the world and yet somehow able to hover above it. Their observations and actions carry enormous weight and garner almost universal respect. In Einstein's case, that meant knowing what was right and what was wrong in the laws that govern our world and in the deeper laws that regulate our universe. It seems unlikely he could have attained what he did without being "the freest man." That freedom brought him to a place of honor and respect as the twentieth century's greatest scientist and one of its great human figures. He was indeed the master of his own destiny, as well as a force in shaping ours.

Mentor and Sounding Board

JOHN ARCHIBALD WHEELER

JOHN ARCHIBALD WHEELER, after notable early contributions to nuclear and particle physics, became a leading figure in the development of general relativity and quantum gravity. He has guided the work of more than fifty doctoral students, beginning at the University of North Carolina, where he taught from 1935 to 1938, then in a long career at Princeton University (1938–76) and at the University of Texas in Austin (1976–86). Among his books are *At Home in the Universe, Journey into Gravity and Spacetime,* and (with Kenneth W. Ford) the 1998 scientific autobiography *Geons, Black Holes, and Quantum Foam: A Life in Physics.* The term "black hole" is his coinage.

When I first met Albert Einstein, in the fall of 1933, I was twenty-two and a freshly baked Ph.D. Einstein was fifty-four and the world's most celebrated scientist. With the brash confidence of youth, I had a clear view of where the frontier of theoretical physics lay at the time—with nuclear physics and with the quantum theory of electrons, positrons, and photons (pair theory, as we called it then). That was the frontier where I wanted to work. Equally clearly, I saw the path that Einstein was then following as a dead end.

The occasion of our meeting was Einstein's first lecture in Princeton, where he had recently arrived to take up his position at the Institute for Advanced Study. For fear of a large crowd, the lecture was not announced publicly. But Eugene Wigner, a Princeton professor (who would later become my close friend), called up Gregory Breit at New York University and invited him to attend. I was a postdoc with Breit at the time, and he invited me to accompany him. We rode the train to Princeton, a trip slightly delayed by its (fortunately harmless) encounter with a farmer's truck, were introduced to Einstein (by H. P. Robertson, as I recall), listened to Einstein's lecture on unified field theory, and, after the usual give-and-take of questions and discussion, returned to New York.

I was impressed by Einstein's stubbornness in pursuing his dream and by the clarity of his presentation—in very acceptable English—but not impressed by the dream itself: a unified field theory of gravitation and electromagnetism. (Einstein, with a bit of self-deprecating humor, had expressed his own doubts about the work he was presenting.) I was convinced then, and remain convinced to this day, that if there is to be a unified theory of physics it must involve more than gravitation and electromagnetism. It must also include quantum entities such as fermions. (Who knows? Maybe the string theorists are on the right track.)

I was rash but, in retrospect, not wrong. Einstein's important work was indeed behind him, and his efforts to unify gravitation and electromagnetism were leading nowhere. But what I still had to wait a few more years to learn was that he retained an astonishing insight into physics and that it was hard to raise a subject in theoretical physics with him that he had not already thought deeply about.

A minisabbatical from the University of North Carolina gave me my first opportunity to get acquainted with Einstein. During my three months at the Institute for Advanced Study, December 1936 to March 1937, I began to appreciate what Einstein had to offer this young theorist searching for the most fruitful problems to work on. For my first lecture at the Institute, I gained encouragement just by his presence. When I arrived early to check out the lecture hall, I found Einstein already seated, the first member of the audience to arrive.

After I joined the Princeton University faculty in 1938, I began to seek out Einstein regularly. We usually met in the second-floor study of his home at 112 Mercer Street rather than at the Institute. He was always gracious, opinionated, lighthearted, and helpful. Although we became close and friendly, we never became close friends. The differences in our age, fame, cultural backgrounds, and worldviews could not be fully bridged. Yet Einstein was my most valued mentor and sounding board for all the years that remained of his life.

In January 1939, Niels Bohr set sail from Denmark to spend several months in Princeton, where he and Einstein could renew their already famous "debates" (friendly discussions, really) about quantum theory. But the news of nuclear fission, which Bohr received shortly before boarding his ship, changed the agenda. I like to say that I feel guilty for my role in preventing these discussions from taking place in anything like the extended and leisurely form that Bohr and Einstein had hoped for. Working with me, Bohr chose to concentrate that winter and early spring on understanding fission. Even though Bohr's office was next to Einstein's in Princeton University's Fine Hall—mine was down the corridor—the two of them spent little time together. (Pending completion of the Institute's first building across town later in 1939, Einstein and other Institute faculty members were housed on the university campus.)

Before long, I myself had a chance to join the quantum debate with Einstein. In 1941, when my gifted graduate student Richard Feynman came up with what he called the path-integral method of quantum theory (and which I renamed the sum-over-histories method), I made an appointment with Einstein and rushed excitedly to his house to tell him about it and get his reaction. The point I wanted to make was that Feynman's method provided a rationale for the probability feature of quantum mechanics that so troubled Einstein. If a particle followed all paths at once to reach its destination (albeit virtual paths)—which is the essence of the sum-over-histories approach—then one doesn't have to face the mystery of how the particle "decides" which path to follow. I explained the idea and then said, "Doesn't this new way of looking at quantum mechanics make you feel that it is completely reasonable to accept the theory?"

Einstein was having none of it. "I still can't believe that the good Lord plays dice," he answered. As Einstein said about himself, he could be as stubborn as a mule. There was no moving him from his conviction that quantum mechanics, at its core, is defective.

The next time I had occasion to consult Einstein on something that Feynman and I were working on was in the late 1940s, after I had returned from my weapons work in World War II. Feynman was visiting Princeton and went with me to Mercer Street. This time the subject was classical physics, so Einstein's allergy to quantum mechanics did not need to come into the discussion. Feynman and I had decided to banish fields from electromagnetic theory, and we found to our delight that the entire theory could indeed be formulated as an "action-at-a-distance" theory without fields, provided we accepted the reality of "advanced" effects—those in which the effect precedes the cause. One of the intriguing conclusions of our work was that in a world of very few particles, time would run backward and forward with almost equal ease, making it commonplace for the future to affect the past, whereas in our actual world of countless trillions of particles, the combined effect of all the distant absorbers of radiation puts a damper on backward-in-time effects, producing the one-way flow of time that we observe.

It was natural to want to talk this over with Einstein. He nodded in agreement and understanding. He had always believed, he said, that the fundamental laws of electromagnetism contained no preference for running forward rather than backward in time. The observed one-way flow

31

of time, he said, is of statistical origin. It comes about, he was convinced, because of the large number of particles in the universe that can interact with one another. Here was Einstein's astonishing intuition at its best. Feynman and I had been through lengthy calculations; Einstein had surmised the result. In fact, thanks to an earlier conversation with Einstein, I already knew about a paper he had written jointly with the Swiss physicist Walter Ritz way back in 1909. It was an unusual paper in two respects: first, it set forth opinions, not demonstrable results; second, the two authors didn't agree. In the paper, Ritz argued that nature's irreversibility (that is, the one-way flow of time) is itself a fundamental principle of nature. Einstein argued instead that "irreversibility rests entirely on probability considerations." The work that Feynman and I did forty years later backed up Einstein's contention.

Despite my ever-growing respect for Einstein and my ever-stronger reliance on his wisdom, I continued to doubt the value of his work on unified field theory. And I was concerned, too, about the number of promising young theorists who wanted to apprentice themselves to Einstein in this work (one-legged physicists, I called them), as well as those to whom general relativity meant juggling subscripts and superscripts without concern for confronting theory and experiment. Accordingly, I was reluctant to get into serious research in gravitation physics myself, because, for me, doing research means guiding students, and I did not want to send students out into the world who weren't broadly trained to work in various branches of theoretical physics and who were not sensitive to the experimental implications of their work—in short, who were unable to stand on two legs.

In the early 1950s, my attitude changed. I saw particle physics headed toward a complex thicket of pions and countless other particles, and I began to sense that there might be more gold in the general relativity mine than had yet been unearthed. To learn, I had to teach, so I was delighted when, in May 1952, Allen Shenstone, the chair of the Princeton Physics Department, approved my request to teach Princeton's first-ever full-year course on relativity in the coming academic year. I was determined to push Einstein's great theory to its limits to see what new insights might be derived.

I didn't discuss my teaching plan right away with Einstein, but once the course got under way I wanted my students to have a chance to encounter the inventor of relativity directly. In the spring of that first year (on May 16, 1953, to be exact), Einstein invited the whole class to tea at Mercer Street. Part of my assignment to the class the preceding day had been to list "three questions that you would like to put to Einstein, with a one-paragraph elaboration of each." When it came to it, some of the students were too tongue-tied to ask any questions, so I had to speak for them. Tea was provided by his stepdaughter Margot Einstein and Helen Dukas, his secretary, and Einstein was a most genial host, fully relaxed even if the students were not.

Among the subjects I wanted to discuss with Einstein at that time was Mach's principle, the idea that inertia arises from the distributed mass throughout the universe. This provocative principle had enjoyed a checkered history since Mach first put it forth early in the twentieth century. Philosophical twaddle, some had called it. I took it seriously, and in 1979 poked fun at those who didn't with an epigram of sorts: "Mystic and murky is the measure many make of the meaning of Mach." Back in 1913, Einstein had also taken it seriously, writing to Mach, "For it necessarily turns out that inertia originates in a kind of interaction between bodies, quite in the sense of your considerations on Newton's pail experiment."* (Newton had pointed out that acceleration seems to be absolute because water in a spinning pail after a time adopts a parabolic surface, quite unlike the flat surface it retains if the pail remains stationary while matter in its vicinity rotates about it.)

But in 1953, Einstein was not enamored of Mach's principle. Surprising both me and my students (for I had told them about Einstein's letter to Mach), he said that he was no longer attracted to Mach's principle. Perhaps, he said, there was nothing in nature corresponding to that principle after all. I had no opportunity then or later to explore with Einstein what had brought about his change of heart.

At that tea, one student was bold enough to ask, "Professor Einstein, what will become of this house when you are no longer living?" Despite his basic humility, Einstein understood well his celebrity status. He answered, as

*The original is in German. For the full text of the letter, see Charles Misner, Kip Thorne, and John Wheeler, *Gravitation* (San Francisco: W. H. Freeman, 1973), pp. 544–45.

nearly as I recall, "This house will never become a place of pilgrimage where the pilgrims come to look at the bones of the saint." And indeed it hasn't. Other Princeton luminaries have lived in the house, but tourists are limited to looking at it from the outside.

Einstein was well aware, too, of his reputation for eccentricity. My secretary, Jackie Fuschini, speaks about her "Einstein sightings" when she was a child. Her mother would take her walking on Nassau Street, perhaps to visit Nill's Colonial Bakery, which also must have been a favorite of Einstein's. On the street or in the bakery, they would occasionally see the famous professor ("the smartest man alive," as Jackie was told), with his hair going every which way, wearing old sneakers with unmatched socks or no socks at all. He responded to greetings with a smile and a nod. And once he invited Jackie's grade-school class to his house—not for tea, just for a visit to shake hands and exchange a few words. So I wasn't the only Princeton teacher to take a class to 112 Mercer Street.

My students' last encounter with Einstein came on April 14, 1954, toward the end of my second year of teaching relativity. Einstein had accepted the students' invitation to speak at a special seminar they had organized. He talked about the early evolution of his thinking, about how he had been troubled from the first about the limitations of special relativity, and about his continuing dissatisfaction with quantum mechanics. The students were thrilled by a presentation that was really a memoir and by Einstein's willingness to respond thoughtfully to their questions. This may have been Einstein's last lecture. He died a year later.

My first paper on relativity was on geons, hypothetical bundles of electromagnetism so concentrated that they are held together by their own gravity (in this paper I also introduced the idea of quantum foam). I completed a draft of the paper in Switzerland in the summer of 1954 and sent a copy to Einstein as soon as I got back to Princeton. Sometime in October, he answered in writing, saying that he had "a pretty bad conscience" about not responding right away, and added, "It is so much easier to understand the other man if he tells you his reasons orally." So I telephoned him to make a date to visit and talk. According to my notes, we had our conversation about geons then and there over the phone—perhaps Einstein suggested that we do so instead of waiting until we could get together.

I should hardly have been surprised to learn that the idea of supercompressed energy had crossed Einstein's mind years earlier. He said he had dismissed the idea as "unnatural." He was prepared to admit, he said, that his equations of relativity allowed for geon solutions of the kind I was exploring, but he doubted the stability of such entities. As I was able to prove several years later, Einstein was right about the instability. (That was not enough reason, however, to abandon the study of geons. Besides photon geons, there could be neutrino geons or even gravity geons—and, who knows, once quantum physics is stirred into the mix, there might even be minuscule geons. Even as transitory entities, I reasoned, geons could be important in the evolution of the universe.)

My geon paper was mostly classical (i.e., nonquantum), but it contained a few remarks about quantum physics, enough to elicit comment from Einstein. He told me once again, as he had so often in the past, that he did not like the probabilistic nature of quantum theory. Nearly fifty years had passed since he had introduced that prototypical quantum entity, the photon (as we now call it). He couldn't stop thinking about, and worrying about, the quantum world that he had helped to bring into being. Now, in my own later years, I find myself pondering quantum theory, too, with one of my favorite questions: Why the quantum? There is something about quantum theory that is rightly more troubling than relativity, something still calling out for a deeper explanation.

That conversation with the seventy-five-year-old Einstein about geons may have been my last with him. He died just six months later, in April 1955. How wonderful it would have been if I had been able to sit and exchange ideas with him in the years ahead, when my students and I were exploring black holes, wormholes, gravitational radiation, and more, extending his own ideas in every direction, and when Charlie Misner, Kip Thorne, and I labored in the early 1970s to assemble what we knew about relativity. Einstein, with his enormous depth of insight, might not have been very surprised by what we were learning, but I like to think that he would have been very interested, maybe a little excited, and surely encouraging. Often I have imagined sitting with Einstein discussing black holes, as the evidence for their reality has accumulated.

My Einstein Suspenders

GEORGE F. SMOOT

GEORGE F. SMOOT received his Ph.D. in physics from MIT in 1970 and was a postdoctoral researcher there before moving in 1971 to the University of California at Berkeley. His group at Lawrence Berkeley National Laboratory conducts experiments observing our galaxy and the cosmic background radiation. The best known of these is COBE (the Cosmic Background Explorer satellite), which has shown that the cosmic background radiation intensity has a wavelength dependence precisely that of a perfectly absorbing body, indicating that it is the relic radiation from the Big Bang. Smoot's honors include, besides the Medal of the Einstein Society, NASA's Medal for Exceptional Science Achievement, the Kilby Award for contributions to science and technology, and the Department of Energy's Ernest Orlando Lawrence Award. He is the author (with Keay Davidson) of a popular book on cosmology entitled *Wrinkles in Time.*

Albert Einstein is such a towering figure that he long ago achieved the status of public icon. Once, needing suspenders for a tuxedo, I went shopping and found a limited choice of patterns: boring geometric, Marilyn Monroe, and Albert Einstein. After much thought, I settled on the last.

Here's one of the ubiquitous anecdotes I heard many years ago and particularly appreciate now. It became a tradition in Einstein's later years for him to grant interviews to the press on his birthday. One year a reporter asked him whether he could imagine having lived a different life. Would he have been happy in another profession? After a moment of reflection, Einstein replied, "I think I would have enjoyed being a plumber." After this remark was reported, the Plumbers and Steamfitters Union, AFL, in Washington, D.C., voted to grant Einstein an honorary membership, and later a New York plumbers' local presented him with a gold-plated set of plumber's tools. Einstein was said to be highly pleased. One day, Einstein's neighbor, a younger physicist, came over and asked to borrow a pipe wrench because his kitchen sink was leaking. Einstein replied, "Sure, if you'll let me help. You don't know how long I have been waiting to use it!" As a do-it-yourselfer myself, I like to picture Einstein with his gold-plated wrench and his legs sticking out from under the sink, dirty water dripping on him as he tries to get the connection properly fitted without skinning his knuckles.

Einstein was not known as a hands-on physicist—that is, an experimentalist—but as a thinker, a theoretician. When I was a neophyte physicist, my primary scientific hero and role model was Enrico Fermi, who was superb both as a theorist and as an experimentalist. From several people I have heard the story of his stunning insouciance on the day in 1942 when the world's first nuclear reactor, which he and his team had built at the University of Chicago, was ready to be tested. Just before it was to go operational for the first time, he called a break for lunch. Only after lunch did the team return and successfully initiate the first sustained nuclear reaction. Fermi went on to Los Alamos as one of the leaders in the Manhattan Project, but he wasn't just a tinkerer. In the 1920s, he and Paul A. M. Dirac worked out the quantum behavior of half-integral spin particles. In this period, Einstein developed the ideas of Satyendranath Bose to understand the statistics of integral spin particles. Fermi was nicknamed "the Pope" by his colleagues, because of his reputation for scientific infallibility.

In the early phase of my career, my research was influenced almost equally by the work of Fermi and of Einstein. There were times, however, when I felt that Einstein was getting more public attention and credit than was warranted, relative to the scientific contribution of others. The press, and thus the public, will often focus on an individual, especially one who captures the imagination and seems accessible. Einstein had an attractively human side, as his plumbing aspirations indicate; perhaps on the day he confessed them to the press he was wistfully hoping for a chance to lead a normal life of independence, such as a plumber might. Fermi, outstanding in both experiment and theory but not as iconic a public figure, seemed like a better role model to me at the time, a sentiment shared by a number of my colleagues. We made much of our scientific lineage, tracing it back through our Ph.D. advisers to

Fermi, and on back to Galileo. It was through this handing down of training, technique, and scientific attitudes that we felt we had become genuine research scientists. Einstein's approach and Einstein's history seemed at odds with this idea of a scientific lineage. His image was that of the outsider, the solitary genius whose startling new ideas burst from an unexpected quarter.

Later, as science and my career advanced and changed, I found my daily life, both research and teaching, more and more directly affected by Einstein's work. Much of Fermi's theoretical efforts were absorbed into the fabric of a larger model of physics. Einstein's relativities, special and general, continue as whole cloth. For many decades, most physicists have treated them as effectively sacrosanct. There were those who were disturbed by the implications and tried to modify or subvert relativity theory; they were regarded as misguided and offtrack. The several early experimental verifications aside, the theories' beauty and intrinsic symmetry alone seemed a powerful indication of their correctness. Ultimately, in physics, observation and experiment are the final arbiters, and in the last decade there has been a shifting of attitude toward general relativity and a growing acceptance of the idea that it will eventually be supplanted by a more advanced theory, much as Einstein's relativity theories supplanted Newtonian physics.

When you are young, you want to learn the work and theory that have preceded you, and then you want to go beyond. As you get older and have done research and taught for a while, you develop an interest in understanding the thought processes of your predecessors in physics and the trial-and-error aspects of their work. You see that rarely does an idea or result leap full-blown from the mind, as Venus rose from the sea. Much more often there are starts and stops, blind alleys, and a lot of plain, dull work before things emerge or the epiphany occurs—prolonged labor before the actual birth. I have often wondered what special abilities and circumstances led Einstein to his break-throughs in the miracle year of 1905. When I taught special relativity to my physics students at Berkeley, I tended, like many of my colleagues, to follow a well-worn path: first, the Michelson-Morley experiment ("The most impor-tant thing that ever happened in Cleveland"), with its null result on the motion of Earth through the so-called luminiferous aether (thought to be the medium carry-ing light waves) and its demonstration that the speed of light is constant. Then the hypothesis put forth by George FitzGerald to account for this result: that lengths contract in the direction of motion. Then the work of Hendrik Lorentz, who produced the formulas that connect space and time in one frame of reference to another moving at a velocity. Then Einstein's revelation of a whole new per-spective, through the transformations we call special rela-tivity. This made a nice logical, pedagogical chain and helped students to understand and accept special relativity as grounded in experiment.

The flaw in this beautiful account was that Einstein had often denied knowing about the work of A. A. Michelson and Edward Morley; his ideas came from thinking about what it would be like to ride along on a beam of light. It would seem that we were misleading the students to the right conclusions. This discrepancy disturbed me, and I finally searched out an obscure report of an interview with Einstein in Japan in which he remarked that he had indeed heard of the Michelson-Morley result before 1905. Why was this remark buried beneath his other widely covered comments that he had come to special relativity by what appeared to be pure thought? Einstein had certainly known of Lorentz's work, and by implication that of Michelson and Morley.

Though much of his other work is an interpretation of observation (the photoelectric effect, atomic and molecular sizes, Brownian motion), Einstein's relativities do in fact both appear to have come in large part from thought and aesthetic considerations. This reliance on thought alone seemed to me, increasingly, to be setting a bad example for budding physicists, especially aspiring theorists, who all seemed to want to be the next Einstein. Of course, I might have been biased, because almost all my own work was experimental and observational, and it was my firm belief that the integrity and power of science came from probing nature, not from divine insight. This was the experimentalist's canon, capital letters and all:

1) Discover an Important Effect or New Thing Never Before Thought Of;

2) Disprove an Important Theory to Show That New Science Is Needed;

3) Confirm a Great New Theory;

4) Disprove a Competitor's Experimental Results—or—

5) At least Confirm a Competitor's Experimental Results!

While it's easy to see the personal-reward priority in the list, all of these items are valuable and essential to the progress of science—indeed, they are the only way to keep the system self-correcting. Appealing to the beauty and purity of a thought to judge the correctness of science is not at all as robust a path to a correct theory. Einstein himself provides examples of this lack of robustness:

1) The cosmological constant, which he famously referred to as "my greatest blunder." He added it to his general relativity equations in order to produce a static universe. At the time, the universe was presumed to be static, but a decade later Edwin Hubble showed that it was expanding. The cosmological constant was seen to be unnecessary—although it has lately been invoked to accommodate what appears to be an accelerated expansion. *Plus ça change, plus c'est la même chose.*

2) In a letter to Max Born (December 4, 1926), Einstein made this famous declaration: "Quantum mechanics is very impressive. But an inner voice tells me that it is not yet the real thing. The theory produces a good deal but hardly brings us closer to the secret of the Old One. I am at all events convinced that He does not play dice." And yet quantum mechanics continues to be validated by experiment.

3) The role of *Der Alte* is summed up in another pair of quotes: "When the solution is simple, God is answering." And this comment, after Arthur Eddington's eclipse

expedition, confirmed the (corrected) prediction of general relativity, to the effect that Einstein would feel sorry for God if the confirmation had not been made, because "the theory is correct."

Aesthetic arguments, while useful as development tools, especially when there are no observations to guide the effort, made me uneasy—seemed a throwback to Greek reasoning about the celestial spheres. More recently, I came to realize that Einstein based special relativity not on pure thought alone but upon a great deal of physical observation and codifying theory—in particular, electromagnetism and the theory of light via James Clerk Maxwell's equations. Einstein was certainly aware of Lorentz's work, but was coming from the Maxwell side, not the Michelson-Morley results. He was reducing these ideas down to two essential postulates added onto the existing physics: (1) the speed of light is constant and independent of the speed of the source or of the observer, and (2) the laws of physics are the same in every inertial frame. From these two postulates and thought experiments, one can derive all the consequences of special relativity, including the Lorentz transformations, time dilation, length contraction, loss of simultaneity, $E = mc^2$, and the lot! Structured in this way, special relativity is a theory of great beauty and one with surprisingly great implications. It was the reinterpretation of special relativity in 1907 by the mathematician Hermann Minkowski that made its calculations straightforward and helped us realize that we live in four dimensions, three of space and one of time, known colloquially as the space-time continuum. This is the starting point for an understanding of general relativity.

In the spring of 2003, I was contacted by Peter Min-kowski, Hermann's nephew, who informed me that I would be the recipient of the 2003 Einstein Medal, to be awarded in Bern in June by the Einstein Society. I was greatly honored: such a named medal is instantly recog-nized, and the previous winners—Stephen Hawking, Ed Witten, and John Wheeler among them—were illustrious. It was all the better that the medal would be awarded in Bern, where Einstein was living when he published his famous 1905 suite of papers. I was then teaching a senior course in relativity and enjoying it more than usual, since I was paying extra attention to the background of Einstein. The chance to see Bern and to think about how Einstein had lived his daily life during the time he was so produc-tive and innovative was exciting to me.

A high point was a private visit to the Einstein Haus at Kramgasse 49, where Einstein lived from 1903 through 1905, when he was developing and publishing those five remarkable papers. On a main street in the center of the city, one floor above a restaurant that spills out onto the sidewalk, the apartment is maintained and being restored by the Einstein Society. Though the ceremony, the talks at the University of Bern, and the dinner were all wonderful, being in Einstein's home and seeing pictures of his family and associates and material about what was going on in physics at the time had the greatest emotional effect on me—an effect all the more heightened because I was allowed to roam through the apartment alone. It seemed a very good one—cozy, with a nice fireplace, hardwood floors, and many architectural details—for a struggling young man with a wife (Mileva Marić) and a new baby

(Hans Albert was born in May 1904), though the family did have to share a bathroom with a family in an adjoining apartment. The living room has two large windows with flower boxes and a good view of the street below, high ceilings, elegant wallpaper, and plenty of room and comfort for his friends and colleagues who gathered there. Einstein, who had hoped for a position at the university, first supported himself and Mileva by a temporary position as a mathematics teacher at the Technical High School in Winterthur. Another temporary position teaching in a private school in Schaffhausen followed. In 1902 he obtained the job as a patent clerk that provided stability and allowed them to rent the apartment. He must have had a burning desire to do physics, what with the distractions of the job and family and the need to finish his Ph.D. thesis. Yet he managed to meet regularly to talk physics with his friends and he found the time to write his papers.

I spent some time going around the city, which is little changed since Einstein's day, taking in the shops and cafés, walking to the university, enjoying Bern, and imagining what Einstein's life might have been like. How and how much did his surroundings affect him? Where and how did he get and develop his ideas? Was it in the quiet time in the patent office or during conversations with his friends, going to lectures at the university, doodling on napkins at a café? Did the pace of life and the intellectual exposure with time to think make it possible? On the weekend, I took a train from Bern to the Alps and hiked above the Lauterbrunnen Valley, across from the Jungfrau, as I guessed Einstein might have done. I wondered if the beauty of nature and the physical monotony of walking had freed his mind to new ways of looking at old things. I found that I was distracted much of the time. But if you prepare yourself well with what is known to be valid, perhaps Einstein was right: careful thought is the way to new understanding.

Beginning in 1905, Einstein embarked on a journey that no one since has equaled: a decade-long run at the cutting edge of physics. This is what we celebrate a century later. I wear my Einstein suspenders with pride.

Einstein, Moe, and Joe

LEON M. LEDERMAN

LEON M. LEDERMAN, director emeritus of the Fermi National Accelerator Laboratory, received the Nobel Prize in physics in 1988 (with Melvin Schwartz and Jack Steinberger) "for the neutrino beam method and the demonstration of the doublet structure of the leptons through the discovery of the muon neutrino." In 1993 he was awarded the Enrico Fermi Prize by President Clinton. He is the author of several books, including (with David Schramm) *From Quarks to the Cosmos: Tools of Discovery*, (with Dick Teresi) *The God Particle: If the Universe Is the Answer, What Is the Question?*, and (with Christopher Hill) *Symmetry and the Beautiful Universe*.

It is difficult to convey—even to the most scientifically oriented of lay readers—the awe one scientist feels for another who has done something truly spectacular. If we examine the Gaussian spectrum of physicists, extending from just-barely-made-Ph.D. all the way to genius, the appreciation of Einstein's achievements only grows, until we get to the (possibly nonexistent) superstar who, now or in the next decade or so, sees a genuine "greatest blunder" buried in the general theory of relativity.

Einstein may be special—so well known through his writings in so many different spheres that the term "legend" is hardly appropriate. Here I want to tell a story and then make a statement about A.E. Telling stories is something I do a lot, after more than thirty years of teaching physics.

Sometime around 1950 a mathematician friend at Princeton asked me if I would like to meet Einstein. At that time, I was a graduate student at Columbia University's Nevis Laboratories, working on its new Synchrocyclotron. Then the most powerful particle accelerator in the world, the machine could accelerate protons to the incredible energy of 400 million electron volts (400 MeV). For scale, the equivalent machine today at Fermilab reaches two trillion electron volts (2 TeV). And so it happened that my best friend from high school, Martin Klein—then a graduate student in theoretical physics at MIT—and I were seated on a bench in Princeton waiting for the Master to pass by with his assistant, Ernst Strauss, who had arranged an introduction. My more-than-fifty-year-old recollection is shaky and would not hold up in any court, but here's how I remember it:

Sure enough, here they come. Einstein has on his usual costume—sweatshirt, baggy pants, sandals. They stop, and Ernst asks him if he would mind meeting some physics graduate students. "No, it will be a pleasure," says Einstein.

We stand, and he asks Martin, "What are you working on?"

"Quantum theory," says Martin.

"Ach! A waste of time!"

52

Einstein then turns to me, and I hasten to say that I am doing experimental research on the properties of pions. These subnuclear particles had been discovered a few years earlier in cosmic rays and were supposed to produce the strong force that holds the atomic nucleus together; the Nevis accelerator was a prolific source of them.

Einstein nods, then shakes his head and says something to the effect that it is already impossible to explain the existence of the electron, so why spend so much effort on these newer particles? He bids us a cheery goodbye, having crushed us both in about thirty seconds. However, we were way up in the clouds. We had met and talked physics with Einstein! The thrill was unimaginable—what he said hadn't mattered at all. Since then, Martin has become a leading scholar in the history of physics and a coeditor of Einstein's papers, and I have helped to discover additional useless fundamental particles, such as neutrinos and quarks.

Why was I not upset by the meeting with Einstein?

This question involves how physicists evaluate major physics achievements, which is clearly different from how laypeople, even science groupies, evaluate them. If we consider a particular discovery or creation—for example, the general theory of relativity—then the appreciation of this seminal achievement will still be driven by history and personality. Physicists recognize that the general theory was uniquely Einstein's. He labored over it for a decade. His drive was not to explain a plethora of experimental results but to express the beauty and simplicity of nature. (His personification of nature was *Der Alte*, the Old One.)

Experiments were of course relevant, and over the decades after the 1916 paper, experiments of awesome precision affirmed that relativity might be a correct theory of gravitation.

So, was that lonely mind influenced? Yes, by Ernst Mach, by James Clerk Maxwell, by mathematical helpers—but in this search for a more profound simplicity in the nature of space, time, and gravity he was very much alone.

Let me place myself somewhere on the bell curve of physicists—say, midpoint—and try to describe how physicists think about Einstein and the very few others who have made major breakthroughs: Newton, Maxwell, Bohr, Schrödinger, Heisenberg, and Dirac. Every one of us has such a list, and my guess is that these names would be included in most. But to me, Newton and Einstein, in true Christmas tree fashion, flash on and off. They were all alone in what they did. Yes, they had guys nearby: Henri Poincaré, Hendrik Lorentz, and Mach for Einstein; Robert Hooke and Gottfried von Leibniz for Newton. But these two were truly far out there, all alone.

The Einstein prejudice, for me, stemmed from my reading, aged about sixteen, *The Evolution of Physics,* a popularization for nonscientists coauthored by Einstein and the Polish physicist Leopold Infeld. The book introduced the theory of relativity but also provided an insight into Einstein's philosophy. What I recall most vividly was its opening metaphor: The authors compared science to a detective story. The way I tell it now, there is a white Ford, a barking dog, a bloody glove, of course a body or two. These and other clues are meticulously recorded, and ultimately the detective (scientist) assembles the suspects and solves the crime, thereby accounting for all the clues.

Here I should record my subjective reaction to other major physics breakthroughs. Somewhere in high school—before 1939—I read about Niels Bohr's use of the concept of quantum energy levels in the structure of the hydrogen atom. Bohr blended a mixture of classical physics and his ad hoc and shocking introduction of discreteness in atomic structure. He also adopted the Planck-Einstein concept of photons—bundles of light energy. The precise wavelengths (colors) of the many spectral lines of the hydrogen atom followed, after some lines of simple algebra. What made teenaged Leon gasp with excitement was the collection of symbols clustering in front of the terms that enumerated the spectral lines. There one found the velocity of light, the charge on the electron, Planck's constant, and assorted two's and pi's.

How could these constants, originating in totally different contexts, come into a description of the hydrogen atom and correctly and precisely give rise to the spectral lines emerging from glowing hydrogen gas? I recall putting the book down and pacing our house, frustrated that there was no one with whom I could converse about this amazing discovery. I had learned an incredibly profound concept about physics: that an idea articulated and composed in the music of mathematics can precisely describe a complex but beautiful piece of nature.

Another graphic example of creative imagination and the profound respect that nature has for mathematics is

Paul Dirac's famous equation describing the electron. Dirac was obsessed by the beauty of equations; his equation for the electron was not only beautiful but also unexpectedly fruitful. In the sense that the square root of four is plus two but also minus two, the equation for the electron predicted two electron-type particles: a negative electron (Dirac's objective) but also a positive electron. Dirac's urge to elegance and beauty had uncovered a revolution in physics: the existence of antimatter. For every particle—electrons, protons, neutrinos, quarks—there must be an antiparticle. What Dirac's epiphany illustrates is the deep influence of the concept of symmetry on the physics of the twentieth century. Because symmetry thrives in mathematics, in arts and architecture, in music and mathematics, its influence in physics not only sparked a revolution in theoretical science but also acted as a unifying connection to the humanities.

Now comes Einstein's year of glory.

It has been pointed out in many places that Einstein's miracle year of 1905 followed several years of discouragement—first with the entire process of being examined for his Ph.D. degree, then with the slow acceptance of his thesis paper, and finally with the need for and the difficulty of finding a job in his chosen field. Sitting as a clerk in the federal patent office in Bern, Einstein—then twenty-six years old—caught fire, and in five stunning papers, all published in 1905, the kid solved three of the most important problems in the physics of his time: the existence and reality of atoms and molecules, the quantum behavior of photons, and a new statement of the principle of inertia, first enunciated by Galileo some three hundred years earlier. Since inertia and relativity are closely connected concepts, the new statement is now referred to as Einstein's special theory of relativity.

By the time I was in college, Einstein's renown was so great that it had to color my judgment as to the depth of his paper on special relativity. But my student days were obsessed with questions: Where did he get this idea? Why Einstein? How could such a simple statement of a concept or a principle have such profound implications?

Einstein was examining patents during the long days and presumably working on physics nights and weekends. Why? He had not been driven by some experimental breakthrough (although there were growing experimentally inspired doubts about the Newtonian worldview) but by an aesthetic and deep physical sense of the accordance of symmetry with nature. Since symmetry is closely associated with beauty and simplicity, we come easily to a belief in Einstein's view of how nature works.

The key word, which we learn in graduate courses but which should be taught in high school science courses, is "invariance." When a physical system is observed from different points of view, or when the system is subject to tortures that only physicists are capable of imagining, it is of intense interest to see what changes and what doesn't change. Does part of the system change? The total energy? The entire system? If nothing changes, the system is invariant. This is nature at its simplest. The system's laws of physics do not care whether observer Joe studies the system while at rest (that is, with the same velocity) or whether Moe, equally adept, speeds by with a huge relative velocity. Moe, the careful physicist, sees Joe and all of Joe's experiments from his (Moe's) viewpoint as he moves past, but Moe also sees the same laws—the same rules. This is true, says Einstein, no matter what the relative velocity. Stated in textbook style, the laws of physics are the same for all observers moving with constant velocity.

This was not a departure from Newtonian science, but Einstein was now also dealing with the phenomena of electricity and magnetism. Maxwell had summarized those experimental laws brilliantly in 1860. The summary of the relevant experiments led to Maxwell's discovery that light was an electromagnetic phenomenon. Combined electric and magnetic forces, vibrating and escaping from their wires into space, traveled at the magnificent speed of 186,000 miles per second. The velocity of light, said Einstein, was a law of physics, the same for all observers! Only in this way could the invariance of both Newtonian systems and Maxwellian systems be respected. So simple! But so profound. The assertions, taken together, constitute the special theory of relativity and bring about a revolution in our concepts of space, time, and energy.

All the confusions and desperate efforts to understand the experiments that were crowding classical physics were swept away by these statements. Who could not love the iconoclast who blew up something called "the luminiferous aether"? Could any of the great scientists of the new century—Poincaré, Lorentz—could any of them have created this idea? Have some fun: raise that issue at the faculty club near the physicists' table and then avoid the flying debris and lurid language as that table erupts.

The special theory combines the two ideas: the velocity of light is the same (invariant) for all observers, and the laws of physics are the same (invariant) for all observers moving at constant velocity. The symmetry and elegance of electromagnetism are thereby preserved—but when these ideas were applied to Newton's mechanics, the world changed. This is Einstein's special theory, and the consequences—economic, technological, and scientific—were as profound as the statement was simple.

The startling thing about special relativity is the engineering applications. We should note that nuclear energy itself is not a consequence of the theory, but there is a plethora of devices that make use of one of the major predictions: that as particles move at velocities approaching the speed of light, there is an increase of mass. Devices whose designs depend upon this effect are large radio frequency amplifiers (klystrons); electron accelerators, used by the thousands in cancer therapy; electron microscopes; high-voltage television tubes; industrial accelerators for sterilization and for control of manufacturing processes, such as thickness measurements; and, most spectacularly, high-energy particle accelerators, which advance our knowledge of the structure of matter and energy. Another ever-increasing application is the use of high-energy beams of electrons to produce "synchrotron light," an intense source of X-rays used to etch silicon elements for microelectronics and to give chemists and biologists graphic photographs of the three-dimensional molecular structure of new materials and new chemicals, and data on DNA and other biological structures. All this from a patent clerk with an attitude.

Although the accumulated contribution of these devices to the gross national product is hundreds of billions of dollars, that all pales into insignificance compared with the revolutionary impact of Einstein's conceptual breakthrough. Much of this hovers around the new, subjective interpretation of time, and it is here that most of us plebeian professors and Nobel laureates can only shake our heads in wonder and gratitude.

When Moe, traveling at a high speed relative to Joe, records the same phenomena as Joe is recording, the numbers are of course different. Joe, for example, locates, say, an electron (a component of the system he is studying) at these coordinates: $x = 6.2$, $y = 9.6$, $z = 27.3$ (all in appropriate units—say, meters). He gives as its velocity $v = 9.6 \times 10^8$ m/s along his x-axis. Moe, looking at the same dumb electron, will have different numbers, because his coordinates—his x's, y's, and z's—will be different. The electron's velocity measured in Moe's lab will be different. If we designate positions and velocity as seen by Joe as x, y, z, and v (along x) and t (time when the measurements were made), Joe's electron coordinates are x, y, z, and t. In Moe's lab let's call his measurements x', y', z', v', and t'.

The laws of physics should not depend on the system or the observer, because there is no way to tell whether Joe or Moe or both are moving. We know only their velocity relative to the system. With a little algebra, we can find the relation of these two sets of coordinates. So far, Newton and all his progeny would be happy. However, Newton would immediately say that $t' = t$—that is, the clocks in Joe's and Moe's lab must read the same time intervals. But in special relativity, the rate of timekeeping may not be the same, and the discrepancy will increase as the relative velocities approach the velocity of light. The new and bizarre aspects of time are the fault of Einstein's equations, which twist and embed time with space, to the despair of the earnest undergraduate.

In the hundred years since the special theory was proposed, this prediction—that clocks, synchronized when, for example, Joe and Moe are at relative rest, run at different rates when Moe revs up his lab and speeds off—has been borne out.

Another story: My Ph.D. thesis experiment, carried out in 1950 (hardly a man is now alive . . .), used a natural clock, the radioactive particle called a muon. Accelerators produce muons at very high velocities, but one can also find muons essentially at rest. At rest, their characteristic lifetime—the length of time it takes some set fraction of your muons to decay—has been carefully measured. When the muons are moving at around 98 percent of the velocity of light, their lifetime is extended fivefold! Gee, if Moe could travel at that speed his lifetime would be about four hundred years!

The catch is, he would not be aware of all that extra longevity until he was able to visit his pal Joe and find that whereas only, say, ten years had passed as far as he was concerned, Joe was now fifty years older. Relative to Moe's clock, Joe's had speeded up fivefold. This is equivalent to Moe's clock, relative to Joe, slowing down to allow him to live to age four hundred, as clocked by envious Joe.

This profound alteration in the nature of time is but

one example of the deep philosophical implications of Einstein's revelations about space and time, the pillars of the world we inhabit. It is for me unimaginable that this sweatshirted shuffler, totally unappreciative of two such promising and handsome grad students, could have had the crystalline clarity of thought to see, discover, compose, invent so much simplicity and beauty in our world.

The True and the Absurd

CHARLES SEIFE

CHARLES SEIFE, a staff writer for *Science* magazine specializing in articles on physics, holds a master's degree in mathematics from Yale and a master's degree in journalism from Columbia. He is the author of *Alpha and Omega: The Search for the Beginning and End of the Universe* and *Zero: The Biography of a Dangerous Idea.*

"You are a smart boy, Einstein, a very smart boy," one of Albert Einstein's professors at the Eidgenössische Technische Hochschule (ETH) is supposed to have said. "But you have one great fault: you do not let yourself be told anything." It was a fault that served Einstein well.

Einstein could not simply accept a theory that was handed to him. It was not in his nature, for he was the jujitsu master of physics. Armed only with intellect and the weapon of the gedankenexperiment—the thought experiment—Einstein had an unparalleled ability to overthrow a theory by using its own power against it. The stronger the theory, the more subtle and dangerous the gedankenexperiments; with his reductio ad absurdums, he laid naked the contradictions in the commonsense picture of the universe.

Einstein began grappling early. When Albert was five years old, a woman came to the Einstein family home in Munich to tutor him. The lessons ended when the five-year-old, in a fit of pique, threw a chair at the hapless teacher. Later on, a teacher at his *Gymnasium* admitted that Albert's attitude unsettled him. According to Einstein biographer Abraham Pais, the teacher complained to Einstein that "you sit there in the back row and smile, and that violates the feeling of respect which a teacher needs from his class."

Einstein's apparent insouciance did nothing to endear him to his elders. Indeed, the reason the young Einstein had to settle for a job as a lowly patent clerk was that he had alienated the above-mentioned ETH professor, Heinrich Weber. Weber handed out assistantships to all the other graduating physics students, but he left Einstein to fend for himself. Yet it was Einstein's questioning of authority that led to the thought experiments that would revolutionize physics.

The first of these came before he began his serious training in physics, and it would eventually lead him to his greatest triumphs. As a teenager, he imagined what it would be like if he were able to travel near or at the speed of light: How would he perceive the universe? Though he didn't realize it at the time, Einstein's flight of fancy exposed a major flaw in the centuries-old Newtonian laws of physics.

As far back as the seventeenth century, physicists had realized that light moved at a finite speed. A Danish astronomer, Ole Rømer, realized that his observations of

Io, one of the moons of Jupiter, were slightly awry; Io was often not in the position it was expected to be in. Moreover, the mismatch between the locations of the moons in the telescope and the moons on paper depended on how close Jupiter was to Earth. Rømer realized that the discrepancy was due to light's finite speed—that it takes some time for light from Jupiter's moons to reach Earth and strike the retina of the observer. As Earth and Jupiter move in their respective orbits, they get closer and farther apart and closer again, so the time it takes for light to traverse the space between them gets shorter and longer and shorter. It's as if someone were playing with your clock, making it run alternately faster and slower—which of course would make it seem as though Jupiter's moons were out of position.

Einstein knew even as a teenager that light moved at a finite speed. Its value is now put at 299,792,458 meters per second. But unlike most scientists of his day, the young Einstein realized that something very odd happens as you move at speeds that approach the speed of light. In his youthful gedankenexperiment, Einstein dreamed he was zooming away from Earth at near–light speed, leaving his native Germany far behind. He imagined looking back at the receding planet and spotting a clock ticking away. And this, he realized, would cause a problem.

If he were stationary, Einstein reasoned, the ticking of the clock would be as it should be. Each time the second hand clicked forward, one more second would have elapsed. But a near-speed-of-light Einstein wouldn't see the same thing. Since he was rapidly getting farther and farther away from Earth, the light from the clock would take more and more time to get to him. The light from the clock's face—the image of the second hand ticking away—would get to Einstein late. It would be as if the clock were ticking slowly; each tick might take two seconds or three seconds or ten seconds or more, depending on how fast he was moving away. And if he actually reached the speed of light, it would seem as if the clock's hands had stopped moving entirely. Only the light from the clock at one given moment would ever be visible to him; light from the clock one second later would always be 299,792,458 meters behind him and would never strike his retinas. From Einstein's point of view, it would be as if all the clocks on Earth had stopped, as if all the people on Earth had frozen motionless in place, as if birds had stopped in midair and fish in midswim. This seemed ridiculous, but it was a consequence of the Newtonian laws of motion and the finite speed of light. If you accepted both of them, then you had to concede that a clock would seem to freeze if you were able to move away from it at the speed of light. Interesting, but not earth-shattering . . . yet.

A few years later, with a more refined gedankenexperiment, Einstein would pit two principles of classical physics against each other again and show that Newton had to be discarded. "Lightning has struck the rails on our railway embankment at two places A and B far distant from each other," he would later write. No longer was the teenage Einstein zooming away from Earth at the speed of light; it was a mature, well-trained physicist who was riding a railway car when lightning suddenly struck the tracks in front of and behind the train. A stationary observer midway between the two lightning strikes would say that they were simultaneous; that the two bolts struck the track at precisely the same instant. Einstein in his railway car stopped midway between the two lightning bolts would agree—the two strikes were simultaneous. But Einstein then set his gedankentrain in motion. If the car were moving quickly from A toward B when the lightning bolts struck, Einstein on the car would see bolt B strike before bolt A. Just as moving away from a clock makes it seem to go more slowly, moving away from a lightning strike lengthens the time it takes for the image of the lightning to reach your retina. A would apparently strike after B, so Einstein on the train would say that the events were not simultaneous. Conversely, if the train were moving in the opposite direction—toward A and away from B—Einstein would perceive bolt A striking before bolt B, so again the lightning strikes wouldn't be simultaneous.

Classical physicists had a set of laws that predicted how objects would move. Trains followed those laws. Clocks followed those laws. Light, too, was supposed to follow those laws. But when Einstein arranged these objects in different configurations and analyzed what would occur in each case, he showed that something very odd happened. Three observers moving in three different ways would perceive the flow of time differently. One of them would think the lightning bolts had struck the rails simultaneously; one would think that lightning bolt A had struck before lightning bolt B, and one would think that lightning bolt B had struck before lightning bolt A. Three different observers saw three different things. What could it mean?

Einstein realized that it meant that the concept of simultaneity had broken down. His gedankenexperiment showed that within certain constraints the order of events is mutable: as you alter your motion through space, you also alter your perception of time. Using eminently Newtonian mechanisms, Einstein showed that the old Newtonian belief in an absolute ordering of events was broken.

With a few more thought experiments, Einstein exposed a number of erroneous assumptions implicit in Newtonian physics. Newton assumed that time was absolute—that each tick of a clock would be the same no matter how an observer was moving. Einstein's thought experiments showed that time was not absolute—that your motion through time depends on your motion through space. Newton assumed that length was absolute, that a meter was a meter was a meter. Einstein's gedankenexperiments showed that length, like time, depends on your frame of reference: two observers moving in different manners can disagree about how long an object is. Physicists had once thought there was no speed limit to an object in the universe; keep applying energy and the object will keep increasing its speed without bound. However, Einstein's gedankenexperiments showed that light was the fastest possible speed achievable by any (ordinary) thing in the universe. And later in his career, he came up with a gedankenexperiment involving a falling elevator that implied that gravity would bend the path of light just as it bends the path of a ball, contrary to what classical physicists had assumed.

The gedankenexperiments pitted classical physical theory against itself. The classical laws were extremely powerful: they made specific predictions about the motion and behavior of objects. But the gedankenexperiments turned that power into a weapon. If light followed Newtonian (and Maxwellian) principles, then the concept of simultaneity and the idea of absolute length and time must be discarded; otherwise the theory would be forced into a contradiction. It was an airtight, logical argument. If classical physics is true, then some of the assumptions of classical physics must break down. There was no escape.

But even Einstein, the undisputed master of gedanken-experiments and reductio ad absurdums, could be defeated. With one of his thought experiments he tried and failed to destroy the second big scientific revolution of the twentieth century: quantum theory. Though he used all his intellectual arsenal against quantum theory (which, ironically, he had helped create with his 1905 paper explaining the photoelectric effect by means of quantized light particles), quantum theory overthrew Einstein. The jujitsu master found a seeming flaw in the theory and used the theory's mathematical power against itself, yet at the end of the bout, quantum mechanics still stood.

By the mid-1920s, Einstein had grown dissatisfied with the newly constructed theory's implications. Though quantum mechanics described the behavior of the microscopic realm with incredible precision, its inherent randomness disgusted Einstein, who was convinced that natural law was beautiful and deterministic rather than ugly and stochastic. He wrote in 1926 to Max Born: "Quantum mechanics is certainly imposing. But an inner voice tells me it is not yet the real thing. The theory says a lot, but does not really bring us any closer to the secret of the Old One. I, at any rate, am convinced that He does not throw dice." So Einstein sought to tear down quantum theory with his usual weapon.

In 1935, Einstein, along with two colleagues, Nathan Rosen and Boris Podolsky, came up with a thought experiment that pitted quantum mechanics against itself. In particular, it used one of the surprising principles that came out of the mathematical formalism of quantum theory, Heisenberg's uncertainty principle, to pose a seeming contradiction.

Heisenberg's uncertainty principle states that it is impossible to know certain pairs of properties of a particle at the same time. For example, if you know precisely where a particle is in space, you automatically know nothing about how fast it's moving—and vice versa. A particle's position and its velocity (or, more precisely, its momentum) are "complementary"; if you gain information about one of these properties, you lose information about the other. This is an absolute dictum of quantum theory—it comes as a direct consequence of the mathematics of quantum mechanics. And it appears in quantum experiments: measure a particle's position with great precision and its velocity smears out, taking a wider range of values than before.

The Einstein-Podolsky-Rosen thought experiment had to do with a particle that decays into two equal-sized pieces that fly off in separate directions at the same speed. The two pieces don't have separate identities; since they are moving away with equal and opposite speeds, if you measure the speed of one, you know the speed of the other. Likewise, if you measure the position of one, you know the position of the other. The two particles, in quantum-mechanical terms, are "entangled."

In the thought experiment, Einstein waited until the two pieces fly a great distance away from each other. Then, two observers—one for each particle—measure the particles in different ways. Observer 1 measures the position of particle 1; observer 2 measures the velocity of particle 2. If observer 1 measures the position with perfect precision, he loses all information about how fast it's moving, thanks to Heisenberg's uncertainty principle. Similarly, observer 2 would know precisely how fast his particle was moving but would know nothing about its position. So far, so good. But then the observers can share notes. Observer 1 has perfect information about his particle's position, which (since the particles are entangled) yields perfect information about particle 2's position. Likewise, observer 2's observation would reveal precisely how fast both particle 1 and particle 2 were moving. After sharing notes, both observers would have perfect knowledge about the particles' positions and velocities at the same moment in time—violating Heisenberg's uncertainty principle.

There was only one way out of this contradiction. If, when observer 1 measures his particle's position, particle 2 somehow "feels" the measurement and smears its velocity out in response, then observer 2 would be unable to measure his particle's smeary velocity with any precision. After sharing notes, the two observers would know precisely where the particles were at a moment in time, but neither of them would have any idea how fast those particles were moving. Heisenberg's uncertainty principle would survive, but at a grievous cost. It makes no sense for two distant particles to instantly "feel" a measurement over a great distance. How can a particle whizzing past the Andromeda galaxy, say, respond instantaneously to a scientist's observation of its twin on Earth? Einstein concluded that it couldn't, that this "spooky action at a distance" made no sense. He was right; it makes no sense. But it's true.

Since the 1970s, scientists have been measuring this spooky action in the laboratory. They look on as one entangled particle senses a measurement of its twin many kilometers away. A closer examination of the properties of entanglement shows that you can't transmit information faster than light via the spooky action, so the phenomenon doesn't violate any principles of relativity—all it does is violate common sense. And that is not a mortal sin for a physics theory. Einstein's thought experiment didn't expose a flaw in quantum theory; the theory remained consistent despite the seemingly nonsensical outcome of the thought experiment. As it turns out, nonsense can be correct.

Even in defeat, Einstein discovered an important physical principle; he discovered a counterintuitive consequence of quantum theory that is now a fundamental part of understanding the mysteries of the subatomic world. The exact mechanism by which entangled particles "conspire" is still not understood, even though many physicists believe they have begun to exorcise the spookiness from the action. The one argument that can defeat Einstein's reductio ad absurdum arguments is that the absurd is sometimes the truth.

Albert Einstein:
A Scientific Reactionary

FRANK J. TIPLER

FRANK J. TIPLER, a professor of mathematical physics at Tulane University, is the author of *The Physics of Immortality*, in which he expounds his Omega Point theory of physical eschatology, and the coauthor, with John D. Barrow, of *The Anthropic Cosmological Principle*.

As a child growing up in rural Alabama, I knew of only two scientists: Albert Einstein and Wernher von Braun. Two immigrant German-Americans. The Jew and the SS officer. But in the 1950s Alabama press, no mention was made of either scientist's politics or ethnic background. Only their science—relativity and rocket science—was discussed. Von Braun was written about far more than Einstein, because von Braun was an Alabamian, having settled in Huntsville as head of the U.S. Army's development team at Redstone Arsenal. Alabama has never been famous for its scientists. (The only famous native Alabama physicist was the accelerator designer Robert J. Van de Graaff.) Inspired by the glowing reports of von Braun's work, I decided at the age of five that I would become a developer of spacecraft. At the age of seven, I sent a fan letter to von

Braun, and he (his office, rather; I'm sure he never bothered with children's fan letters) sent me a photo of a WAC Corporal (an early version of the Jupiter-C rocket) and his portrait. Both photos were inserted inside a silver frame given me for that purpose by my grandmother and placed in a prominent position near my bed.

In grammar school I made models of the von Braun space station, multistage orbital rocket, and circumlunar rocket and hung them from the ceiling of my room; these models were widely available in the 1950s, because Walt Disney had based several "Tomorrowland" episodes of his weekly television program on von Braun's early ideas. I organized an astronomy club, at which we discussed these ideas. I began to read elementary physics textbooks, and when these became too difficult (which was fairly soon), I checked science fiction books—usually those of the space travel genre—out of the public library. These were mainly stories of interplanetary spaceflight using chemical rockets—*Tom Swift and His Rocket Ship* was one memorable title—but a few books were about traveling to the stars. It soon became my goal to do von Braun one better: he would perfect interplanetary rockets and I would devote my adult life to perfecting interstellar manned vehicles.

In high school I read Robert Heinlein's *Time for the Stars* and became aware that Einstein's relativity theory was the great barrier to interstellar travel. Heinlein's story emphasized the speed-of-light constraint: Nothing can go faster than light. Heinlein's solution was simply to postulate that some future physicist would find a theory supplanting relativity, a new theory in which it is discovered that the speed of light can be exceeded after all. This seemed plausible: Einstein's theory had replaced Newton's, so why shouldn't Einstein's theory be replaced in turn by Professor X's? I hoped in my heart of hearts that "X" would be "Tipler."

I learned from Heinlein's books that MIT was the place to go if you wanted to get the best education in physics, and I clearly needed the best education in physics if I wanted to figure out a way around that pesky speed-of-light barrier. So I applied to MIT and was accepted, probably not on my own merits but because I was from Alabama. Very few Southerners applied to MIT, and the admissions office wanted a few students from outside the Northeast.

MIT was a rude awakening. Not only was I taught the mathematics of physics but I was also presented with the enormous amount of experimental evidence for relativity—in particular, the evidence supporting the speed-of-light barrier. I would have despaired and abandoned my dream of interstellar flight had not a fellow physics major pointed out, in my sophomore year, a paper by the logician Kurt Gödel, a close friend of Einstein's at the Institute for Advanced Study in Princeton.

Gödel's paper was a revelation. It described a way around the speed-of-light barrier using Einstein's own general theory of relativity. Relativity could be used to overcome relativity! Gödel had found a new solution to Einstein's equations—a solution that represented a rotating universe. Gödel demonstrated that in such a universe it was possible, by means of a rocket that traversed a certain trajectory, to travel in a closed path in time; that is, one could travel to a distant star, taking thousands of years if necessary, and then return to Earth via the Gödel path. One would return to Earth shortly after one had left it. In effect, faster-than-light speed was possible in the Gödel universe even though the interstellar vehicle never exceeded the speed of light locally, and so Einstein's speed-of-light barrier was never breached. Soon after becoming aware of Gödel's paper, I read a paper by Einstein himself on Gödel's rotating universe. Einstein wrote that he thought Gödel's mathematics impeccable, that he had wondered about the possibility Gödel had raised when he (Einstein) was developing his general relativity theory, and that the question of whether such a rotating universe existed or was possible could be established only by experiment.

In my sophomore year I had not understood Gödel's paper very well, but I understood enough to realize that I could accomplish my goal of interstellar travel only if I became an expert in Einstein's general theory of relativity. I also realized that to be a radical in science, one needed to be a conservative; that is, revolutions in physics are accomplished not by deliberately trying to overthrow known physics but by thinking deeply about the full implications of the laws of physics that we believe we know. I vowed to investigate the implications of the Gödel idea. The faster-than-light effect was associated with rotation in general relativity. We certainly can't start the universe rotating if it isn't already. But I wondered whether a smaller rotating body would show a similar effect. In my junior year at MIT, Stephen Hawking and Roger Penrose published their singularity theorems, and these theorems were proved using techniques designed to analyze the weird time properties of the Gödel universe. I knew that I had to master these techniques. I had become a student of Einstein's general theory of relativity.

I also began to study Einstein's approach to physics, his research strategy. In particular, I wondered what had led Einstein to develop his theory of special relativity. I found the answer in his "Autobiographical Notes." Einstein wrote that at the age of sixteen (!) he realized that there was an inconsistency between the two fundamental theories of the nineteenth century, James Clerk Maxwell's theory of electromagnetic radiation and Isaac Newton's mechanics of particles. According to Maxwell's theory, light is a form of electromagnetic radiation and a form of wave motion. The waves of the sea, as they approach the shore, are the most familiar form of wave motion. According to Newton's mechanics, it is possible to move at any speed, so Einstein imagined moving at the speed of light alongside a wave of light. Surfers move at the speed of water waves, so this seems a reasonable thing to imagine. From the surfer's vantage point on top of the wave, the wave appears stationary, so Einstein reasoned that if he moved alongside a light wave at the speed of the wave, the light wave should appear stationary to him, just as the waves of the sea appear stationary to the surfer. Mathematically, this means that there should be a solution to Maxwell's equations describing this stationary wave. (There are solutions to the water-wave equations for the surfer's stationary wave.)

But, as Einstein realized, there are no such stationary

solutions to Maxwell's equations. As Einstein put it in his "Autobiograpical Notes": "I should observe such a beam of light as a spatially oscillatory electromagnetic field at rest. However, there seems to be no such thing, whether on the basis of experience or according to Maxwell's equations." These equations seemed to be saying that it was impossible to move as fast as a light wave. Since Newton's mechanics said that it *is* possible to move at such a speed, Einstein had discovered a fundamental contradiction between the two theories of his day. Either Maxwell's theory or Newton's theory, or both, had to be wrong. At least one of the theories had to be modified.

Einstein resolved the inconsistency in 1905, when he was twenty-six years old. He later remarked that once he realized that Newton's idea of absolute time was suspect, he was able to work out within six weeks how to modify Newton's mechanics to make it consistent with Maxwell's equations. Because there are revolutionary implications to Einstein's mechanics—$E = mc^2$ being the most familiar—it is not often realized how profoundly conservative Einstein's innovation actually was. It was a minimal modification of the fundamental physics equations of his day. Maxwell's equations have a fundamental speed—the speed of light—built into them in an essential way. Removing this speed would require a major reworking of the equations. By comparison, it was a trivial matter to change Newton's mechanics to include this speed limit, using light signals to coordinate the time measurements of separated clocks (an idea Einstein picked up from his day job as a Swiss patent examiner). Changing Maxwell's equations to make them consistent with Newtonian mechanics would almost certainly have ruined their agreement with experiment, whereas the changes Einstein made in Newton's mechanics would show up only at speeds comparable with that of light. So introducing the speed of light into particle mechanics required no detailed checking of experiments—and Einstein made no such checks. His paper describing relativity had no references. It didn't need them for such a minor change to the equations of physics.

Once Einstein had established that particle mechanics had a fundamental speed limit, the speed of light, it was obvious that Newton's theory of gravitation would also need to be modified, because this theory required gravitational effects to travel with infinite speed. There is no speed limit in Newton's law of attraction; the gravitational effect of moving a rock here on Earth would be felt, in principle, everywhere in the universe instantly. Einstein succeeded in creating a new theory of gravity—the Einstein equations I've devoted my life to studying—by 1917. His theory of gravity, usually called general relativity, is often considered a revolutionary way of thinking about gravity, because Newtonian gravity is a force, whereas in Einstein's theory gravity is the curvature of space and time. But general relativity, too, was really a *conservative* modification of the existing Newtonian gravity theory. The great French mathematician Elie Cartan would prove in the 1920s that Newtonian gravity is not really a force but instead a manifestation of the curvature of time! This follows from the fact that the "force" of gravity acting on a particle is proportional to the mass of that particle. This in turn means that the path a particle follows in a gravitational field does not depend on the particle's mass, since the mass cancels out on both sides of Newton's second law of motion, $F = ma$. This cancellation means that one can regard the path of a particle in a gravitational field as following a "curvature" of either space or time, just as trucks, cars, and bicycles follow the curvature of a highway. The path followed by trucks, cars, and bicycles doesn't depend on how massive the moving vehicle is. Cartan showed that Newton's equation for the generation of gravitational fields could be obtained by assuming that only time is curved. But why should only time be curved? Einstein's theory of gravity allowed both space and time to be curved and showed how the curvature of each was joined together. What could be more natural?

Most physicists now recognize that Einstein's theory of relativity is not a revolutionary theory at all but a completion of classical physics. Einstein's most subtle biographer, Abraham Pais, has conceded this, but also maintained that Einstein's invention of quantum mechanics, in his 1905 paper on the photoelectric effect, was still revolutionary.

I disagree. Einstein's invention of quantum mechanics was, once again, a conservative innovation—conservative in the traditional sense of *preserving* the classical structure of Newtonian physics. Einstein began his paper on the photoelectric effect by analyzing the energy distribution formula for light inside a black container of a certain fixed temperature, a formula obtained five years earlier by Max Planck. He pointed out that Planck's formula had numerous parallels with the energy distribution of air molecules bouncing around inside a similar container of fixed temperature, a formula obtained some twenty years earlier by Ludwig Boltzmann. So similar were these formulas that Einstein concluded that light must consist, like the apparently continuously fluid air, of tiny particles, which would later be called photons. Einstein ended his paper by pointing out that if, indeed, light were made up of such particles, then ultraviolet light hitting a metal surface would dislodge electrons just as the cue ball on the pool table can dislodge a single ball from a racked-up set of billiard balls. Furthermore, the maximum energy of the billiard ball that leaves the collection of balls is proportional to the energy of the cue ball that hits the set of balls. An electron dislodged in this manner from a metal surface is called a photoelectron ("photo" means "light"), and this relationship between the energies of the incident photon and the emitted electron is called the photoelectric equation. For obtaining this equation, Einstein was later awarded the Nobel Prize.

Einstein is supposed to have written a letter to a friend of his in which he called this paper on the photoelectric effect "very revolutionary." But I've noticed that the historians who cite this letter have never seen the letter itself; they all cite the same secondary source. I doubt that Einstein believed it was a revolutionary paper. It was actually a very reactionary paper. By arguing that light consisted of particles, Einstein was returning to a theory of light that was believed refuted in the early nineteenth century, when numerous experiments had been done proving that light was a wave phenomenon. Remember Maxwell's equations?

These in fact assumed that light was a wave phenomenon. Indeed, light *is* a wave phenomenon, and physicists at the end of the nineteenth century were attempting to show that all particles were really a form of condensed light. That is, the idea that an atom—a particle—was a fundamental entity in nature was a very old-fashioned idea by the end of the nineteenth century. By proposing that light consisted of particles, Einstein was trying to turn back the clock.

Einstein was correct in saying that light consists of photons, but the nineteenth-century physicists were themselves correct in thinking that light consisted of waves. The error all physicists before Einstein made was thinking in either/or terms. They assumed that light had to be either particles or waves, ignoring the possibility that light could be both. It was this possibility that Einstein established as fact.

The possibility that light—indeed, every object in the universe, including people and planets—should be thought of as being simultaneously particles and waves should have been obvious to all physicists long before Einstein. It had been established by 1850 that Newtonian mechanics in its most powerful mathematical form, known as Hamilton-Jacobi theory, required that everything in existence be simultaneously both particles and waves. But Newtonian mechanics was not taken seriously in its most powerful mathematical form until Einstein forced physicists to reconsider. In 1926, the Austrian physicist Erwin Schrödinger proved that a mathematical problem with the Hamilton-Jacobi equation could be overcome by the addition of a new term. The resulting equation is today called Schrödinger's equation, and it forms the foundation of modern quantum mechanics. In other words, "conserving" classical mechanics leads necessarily to quantum mechanics. As Steven Weinberg and Freeman Dyson have recently remarked, all achievements in fundamental physics since Einstein and Schrödinger have, once again, been conservative innovations. Dyson has further pointed out that in Einstein's later years, when he deserted his conservative approach to physics for the revolutionary one of trying to develop a unified field theory by pure mental intuition, he failed as a physicist.

For me, Einstein is today what he was for me when I was young: the role model par excellence. I hope to learn from both his successes and his failures. I believe that his successes came from his conservative approach to physics and his failures from his deliberate attempt in later life to revolutionize physics. The great Einstein, the foremost physicist of the twentieth century, was a scientific reactionary!

Helen Dukas:
Einstein's Compass

GEORGE DYSON

> GEORGE DYSON is an author, designer, and historian of technology whose interests have ranged from the history of the Aleut kayak (*Baidarka*) to the evolution of digital computing and telecommunications (*Darwin Among the Machines*) and nuclear bomb–propelled space exploration (*Project Orion*). His early life and work, contrasted with that of his father, the physicist Freeman Dyson, was the subject of Kenneth Brower's 1978 book *The Starship and the Canoe*. He lives in Bellingham, Washington, and divides his time between building boats and writing books.

Albert Einstein remembers being five years old, and sick in bed, when he was given a magnetic compass by his father, sparking a fascination with physics that governed his existence for the next seventy-one years. I was seven years old, and my sister Esther eight, when Helen Dukas, who had been Einstein's personal secretary since 1928 and his literary executor since his death in 1955, began making regular weekly visits to babysit for us and a growing brood of younger sisters at the Dyson household on Battle Road in Princeton. Helen occasionally brought something from the stockpile of games and puzzles that had accumulated in the Einstein household—sample copies that had been sent to Professor Einstein and now ended up with us.

Like Einstein, I remember being given a small magnetic compass, but it failed to have any effect. To me it was just a compass, something to find your way north or south if you got lost in the woods. I did not wonder, Why does it point north? or Why are there electromagnetic fields? I just wondered, Why would I ever get lost in the woods?

The two dozen permanent faculty and two hundred or so temporary visitors of the Institute for Advanced Study occupy a six-hundred-acre campus, of which five hundred acres remains a private nature reserve. My father, the physicist Freeman Dyson, joined the Institute's faculty in 1953, the year I was born, and I spent much of my childhood in the Institute woods. Princeton, New Jersey, is a spacious, forested community—preserved by zoning laws and wealth—but nowhere is the sense of space as luxurious as in the Institute woods. The woods and Albert Einstein are the Institute's two most enduring icons, reminding the world that the Institute exists simply to give a select group of individuals time and room to think. "What could be wiser than to give people who can think the leisure in which to do it?" economist Walter W. Stewart remarked to its founding director, Abraham Flexner, in 1939, the year that Fuld Hall, the Institute's headquarters, was built at the edge of the woods.

The woods are a refuge for wildlife, a haven for scholars, and an irresistible wilderness for kids. The Institute's landscape—roughly one square mile, bounded by Olden Farm, the Princeton Battlefield, Springdale Golf Course, and the Delaware & Raritan Canal—was imprinted on the children who grew up there as indelibly as the details of Christopher Robin's Hundred Acre Wood on the endpapers of *Winnie-the-Pooh*.

The Institute's Christopher Robin was the topologist Oswald Veblen, nephew of Thorstein Veblen, who came to Princeton University in 1905, remaining in Princeton for life, except for stints at the Aberdeen Proving Ground during World Wars I and II and summers at his retreat in the Maine woods. An indefatigable outdoorsman as well as an administrator, he proposed what would later become the Institute for Advanced Study in a letter to the Rockefeller Foundation's Simon Flexner on February 23, 1924. "The way to make another step forward," argued Veblen, "is to found and endow a Mathematical Institute. The physical equipment of such an institute would be very simple: a library, a few offices, and lecture rooms, and a small amount of apparatus such as computing machines."

Flexner wrote back to Veblen, "I wish that sometime you might speak with my brother, Mr. Abraham Flexner, of the General Education Board." Six years later, when Abraham Flexner, then retired from the Rockefeller Foundation, secured funding from the Newark merchant Louis Bamberger and his sister Caroline (Mrs. Felix Fuld) to found a new institute, Veblen was the first professor brought on board. Veblen used his new position to expand the bounds of mathematics in America. He also set the boundaries of the Institute woods.

"So far as I know, there is no educational institution in the United States which has not in the beginning made the mistake of acquiring too little rather than too much land," Veblen wrote to Abraham Flexner on April 12, 1934, urging the acquisition of "a sufficiently large plot of land, which would thus be kept free from objectionable intruders." Veblen tramped through the woods and fields of Princeton, drove a series of tough bargains with Depression-strapped landowners, and assembled the parcels that make up the refuge, now surrounded by development on all sides. In 1959, J. Robert Oppenheimer wrote to Veblen asking permission to change the name of one of the Institute's roads from Portico to Veblen Lane. A notation in the margin of the letter summarizes Veblen's reply: "Said no. Would rather wait until dead."

Veblen was determined to assemble a first-rate school of mathematics at the Institute, building on what he had already achieved at Princeton University, by methodically pursuing the brightest minds in the field. High salaries, no teaching responsibilities, lifetime tenure, and offices with high ceilings and fireplaces were among the lures. Flexner deferred to Veblen on the subject of whom to recruit, explaining to the trustees in 1938 that "Mathematicians, like cows in the dark, all look alike to me." Flexner did, however, make overtures to Einstein, and when Einstein accepted Flexner's invitation he became professor number two. By then, Einstein was not only a prized scientist and public figure, he was a prized refugee. The Institute would take a leading role during the 1930s in harboring those fleeing the ravages of Europe, taking in as many scholars as it could. Veblen assumed the chairmanship of the Rockefeller Foundation's Emergency Committee for Displaced German Scholars, established to counter the twin misfortunes of anti-Semitism in Europe and a depression in the United States. An invitation to the Institute in Princeton was a lifeline to those in danger of being submerged in the descending darkness, and it was the opportunity to take the lead in this mission, in partnership with the Rockefellers, that helped to draw Einstein into signing up with the new institute for life.

Flexner suffered from founder's syndrome, in that he held his new faculty in an uncomfortably tight grip. He saw a firm directorship as necessary to avoid "dull and increasingly frequent meetings of committees, groups, or the faculty itself. Once started, this tendency toward organization and formal consultation could never be stopped." Einstein (and Helen Dukas, too) found Flexner overbearing, and there was tension on both sides. "To build a great institute we have to have sooner or later great people who are different from the ordinary, witness, for example, Einstein, who has done a number of foolish things since going to Europe," Flexner wrote to founding trustee Herbert Maass in 1933. "Of course, I do not allow them to disturb me in the least, for I know that, when he reaches Princeton, I shall contrive to manage him and his wife."

Einstein, however, was using Flexner and the Institute for his own purposes as much as Flexner was using Einstein for his. Princeton was one of the most conservative communities in America, and Einstein doubtless realized that for all its stuffiness and pretensions (and a lingering quota against Jews at Princeton University), this sanctuary in the heart of the East Coast establishment was a good place for a rebellious nonconformist to exert influence from within. Einstein eventually helped lead a faculty rebellion against Flexner, and the director was forced into retreat, replaced by Frank Aydelotte, the former president of Swarthmore, who had joined the Society of Friends in 1939 and whose diplomacy was invaluable both within the Institute and beyond. As a member of the joint Anglo-American commission to Palestine in 1946, Aydelotte helped found the state of Israel, which Einstein would later be asked to lead.

The Einsteins, with Helen Dukas, arrived in Princeton in October of 1933. Helen would remain in Princeton as part of the Einstein household (and manager of the household after the death of Elsa Einstein in 1936) until her death in 1982. Though overshadowed by her employer, she possessed a phenomenal intellect in her own right. "Helen could remember infallibly who had written what when, who needed an answer and who didn't, who was an earnest seeker after truth, and who was a journalistic pest," my father recalled at her memorial, adding that her presence had allowed Einstein "to live the life of an absent-minded professor; she kept to herself the tiresome details that he wanted to forget, and she reminded him of the important things he wanted to remember." To the rest of the world, Einstein achieved immortality through his science, his humanity, and the celebrity he enjoyed while alive. To friends and neighbors in Princeton, Einstein achieved immortality through Helen Dukas. My sisters and I were too young to have known Einstein, but Helen's weekly visits brought him back to life for us.

Helen was born on October 17, 1896, and grew up in Freiburg as one of seven children. She left school to help take care of her siblings after her mother died in 1909, and her first job was as a kindergarten teacher in 1919. There were no children in the Einstein household in Princeton, and her own relatives had remained overseas. Our house was about halfway between the Institute and the Einstein house on Mercer Street, we had no grandparents in America, and Helen had no grandchildren—so everything seemed to fit. Like Mary Poppins, she just showed up one day, having evidently chosen us as much as we chose her.

Helen was stern but kind. She loved to talk—especially in German—but she also knew how to make an impact with just a few well-chosen words. Her knowledge was encyclopedic: almost everything that transpired between the world and Professor Einstein had flowed through her and was mentally cataloged en route. She performed the essential functions of the search engines we take for granted today.

On one particularly memorable dark winter afternoon, she made me start reading my first grown-up book. I had been bouncing around on the large, green, spring-loaded Naugahyde recliner in my father's study, bored and generally making her life difficult, until finally she suggested, emphatically, "Why don't you read something?" To which I answered, "There's nothing to read!" The study was, of course, full of books, but not children's books. Helen went to the bookshelves, picked out a book, and said, "This one. Read this!" The book was Thor Heyerdahl's *Kon-Tiki,* and to me this became *the* book. "Just occasionally you find yourself in an odd situation," explained Heyerdahl in the first sentence. "You get into it by degrees and in the most natural way but, when you are right in the midst of it, you are suddenly astonished and ask yourself how in the world it all came about." I was hooked—suddenly transported out of the green recliner and into the balsa forests of Peru, off into the trade winds, and among the islands of the South Seas. The idea that you could lash a raft together with rope and sail it across the Pacific Ocean became a guiding beacon in my life, much as Einstein's life had been guided by that first childhood encounter with a magnetic field.

I began lashing stuff together—at first, bamboo towers from the bamboo growing along the Oppenheimers' fence—and have been lashing things together ever since. After leaving Princeton at age sixteen, I went to live in a tree house lashed together ninety-five feet up in a Douglas fir on the coast of British Columbia, where I began to study, and build, the kayaks (or baidarkas, in Russian nomenclature) of the Aleuts, who have lashed together some of the most sophisticated watercraft the world has ever seen. Although I was trying to get as far away from Princeton and its institutions as I could, I ended up doing science and history, and in my own small way re-creating Abraham Flexner's vision of an intellectual paradise in the woods.

Did Helen Dukas have premonitions of this? She was an astute judge of character, of families, and of siblings, and she may well have sensed that I would remain in the shadow of my father and my sister (Esther being a more difficult act to follow than Freeman) unless I went off and did something entirely different, such as building Aleut kayaks in the British Columbia woods. My first kayak, built in my Battle Road bedroom, followed a few years after my reading of *Kon-Tiki* and was launched into the Delaware & Raritan Canal near the Institute's swinging bridge. Soon afterward I left Princeton, and it would be a long time, and many kayaks, before I would return.

Just as Helen Dukas managed to define my life's trajectory simply by handing me the right book, I believe she guided Einstein's life by a series of subtle hints. The day-to-day routine of saying, "Here, read this," or "I think we should answer this letter" (while discarding others), concealed a deep understanding on her part of Einstein's place in the world. She allowed Einstein to be Einstein. Her instincts were as infallible and straightforward as a magnetic compass, and Einstein needed this.

I saw Helen Dukas for the last time twenty years after reading *Kon-Tiki* and two years before she died. I was back in Princeton for the first time in many years, fresh from a four-month voyage off the coast of southeast Alaska and British Columbia, retracing the route of eighteenth-century Russian and Aleut sea otter hunters with a fleet of six three-hatch baidarkas that a group of friends and I had built. This was my Thor Heyerdahl period, and I was invited to give a slide show in the round lecture hall in the

Institute's new social science building, the first of several modern buildings that have now sprouted up around Fuld Hall. Helen Dukas and Margot Einstein, Einstein's step-daughter, were in the front row for my talk. I wonder whether Helen noticed that on our voyage we had carried no compasses, navigating the coastline by instinct in the way she was still navigating the Einstein Archive without an electronic database, the way children still played in the Institute woods. Did she remember giving me Heyerdahl's book?

Our kayaks had fan-shaped, downwind sails with which we captured the fjord-funneled southeasters and westerlies of the Inside Passage, just as the *Kon-Tiki* had captured the South Pacific trade winds. The slides showed the latest step in a voyage that had begun when I carried my first kayak through the Institute woods to the Delaware & Raritan Canal, intending to paddle as far away from Princeton and its physicists as I could.

Helen and Margot came up to me afterward, older but unchanged. Margot, with the same brilliant sparkle in her eyes that I remembered from my childhood, told me what Helen had probably known all along but I had never expected to hear: "I wish Uncle Albert could have seen this!"

My Three Einsteins

COREY S. POWELL

COREY S. POWELL, a senior editor of *Discover* magazine and an adjunct professor of science writing at New York University, is the author of *God in the Equation: How Einstein Transformed Religion.*

A half century after he died, Albert Einstein still knows how to make an entrance. He drops in unexpectedly when I take out the trash: a momentary glance up at the night sky becomes a vertiginous vision of fusion-powered stars, their bulks held together by the curvature of space-time, their light emitted at a steady 186,282 miles per second. He leaps out from among the sun-bleached rocks when I visit Mount Wilson in California, where Edwin Hubble first saw that the universe is expanding and so transformed the general theory of relativity into a road map of the origin and fate of the cosmos. And he greets me from the faint, Xeroxed papers of the Einstein Archives at the Institute for Advanced Study in Princeton, his words still fresh and vibrant in letters to Franklin Roosevelt, Sigmund Freud, Bertrand Russell, adulatory children, even to cranks wishing to debunk his theories.

Over the years, these visitations have consolidated into

a portrait of my Einstein—or, more precisely, my three Einsteins, related but distinct aspects of the man, which I envision nested inside one another like Aristotle's heavenly spheres. The symbolic Einstein touches me through his seismic influence on popular culture; the scientific Einstein reaches me through his serpentine formulas and theories; the philosophical Einstein reaches deepest into my heart, challenging my notions of beauty and spirituality. What ties these together is Einstein's miraculous gift for reckless invention. In his public proclamations, his theorizing, and his religious ruminations, he cast a piercing gaze on existing formulations, rejected existing ideas, and freely redefined terms—space and time, pacifism, God— in search of deeper meanings.

The symbolic Einstein offered me his most pointed lessons while I was growing up, just as he has for millions of other academic-minded kids over the past eight decades. Who doesn't know the stories? Einstein famously (if not actually) started out a "slow" child but grew up to become a gentle genius. Einstein was so far ahead of his time, so out of step with his colleagues, that he had to take a menial job at the Swiss patent office while he hammered away at the mysteries of $E = mc^2$. Einstein encouraged the development of the atomic bomb, then spent the late years of his life arguing for peace and international cooperation. He was an otherworldly presence, visually signified by his mane of untamable hair, who nevertheless uttered deliberately accessible epigrams: "God is subtle, but He is not malicious," or "To punish me for my contempt for authority, Fate has made me an authority myself."

It hardly matters that many aspects of Einstein's pop biography verge on caricature. The messages they convey are valid all the same. This Einstein taught me that great achievement is inextricably tied to a healthy dose of disrespect for mainstream belief. For me, Einstein was a kind of physics hippie, a man whose creativity was inseparable from his refusal to play by the rules of academia and buy into its comfortable certainties. He reminds me of Bob Dylan kicking out an electrifying "Like a Rolling Stone," or John Lennon embracing guitar feedback and Yoko Ono's abrasive Fluxus art. Einstein could easily have compromised, working more on the applied side of physics and taking on teaching duties. Instead he chose a line of work that allowed his thoughts to hum freely until they spun out the song of special relativity.

I still marvel at the diligent vigor that enabled him to remain true to his freethinker conventions even as his place in the world, and the world around him, changed. His fame built steadily after the publication of special relativity in 1905 and accelerated sharply when he unleashed the general theory of relativity in 1915. Then came the crescendo, when the prominent English physicist Arthur Eddington examined observations collected during a 1919 solar eclipse and declared that the sun's gravity bent the light of nearby stars in exactly the manner Einstein predicted. Suddenly Einstein went from the physics journals to the front pages of the world's newspapers and morphed into science's first modern media star.

The adulation changed him, but not in narcissistic ways. He still followed a resolutely independent path through the landscape of physics, seeking a single theory of all the laws of nature. Few followed his lead, and his many published attempts at a unified field theory proved to be frustrating dead ends. He persevered all the same, reportedly calling for a notebook on his deathbed in the hope that a final flash of inspiration might complete the project of his last thirty years.

Even as Einstein's scientific inspiration dimmed, fame exposed another aspect of his greatness: a profound understanding of the responsibility that came with celebrity. He was keenly aware that he had become the public face of science, a role he treated with seriously playful irreverence. The genial "Uncle Albert" persona undercut the stereotype of the scientist as a coldhearted materialist. (Think about how many photographs of Einstein on the bicycle or Einstein sticking out his tongue still stand watch over college dorm rooms.) Those famous quotes citing God did the same, in a more penetrating way. I interpret Einstein's use of that word as a symbolic act as much as a theological one. He evidently understood that a science that ignores or seemingly refutes religion would never be fully satisfying to the public—not even to himself.

In politics, too, Einstein carefully evaluated the interplay between his authority and his contempt for authority. He had always been an ardent antinationalist and pacifist, vehemently opposed to World War I and appalled by his many German colleagues who supported it. Now he clung to these ideals but recognized the danger of blind adherence to ideology, even the idealistic ideology of pacifism. In a 1931 talk at the California Institute of Technology, he explained how he had reinvented the word: "I am not only a pacifist but a militant pacifist. I am willing to fight for peace. . . . Is it not better for a man to die for a cause in which he believes, such as peace, than to suffer for a cause in which he does not believe, such as war?" When the threat of a Nazi atomic bomb seemed real, he signed the letter drafted by physicist Leo Szilard urging President Roosevelt to begin an American atomic bomb project. Yet he stuck to his core beliefs, arguing after the war for disarmament and international government to preserve the peace.

Who in the generations of scientists after Einstein has shown such a clear understanding of what to do with the bully pulpit of public acclaim? Look at the most famous living scientists—Stephen Hawking, for example, or James Watson. They have all contributed wonderful ideas to the world, but few interact meaningfully with today's major political and social issues.

Beneath the symbolic Einstein lies the scientific Einstein—the man whose theoretical breakthroughs justified the fame and all that flowed from it. Here, too, he was more than just a superficial radical. Physicist Banesh Hoffman, Einstein's collaborator and biographer, aptly described him as "creator and rebel." Einstein destroyed classical certainty, but only so he could uncover a deeper kind of certainty. Isaac Newton had envisioned a universe built on absolute space, an invisible metric against which all motion can be measured. Einstein replaced Newton's universe with one built on absolute *law*, meaning that the speed of light and other fundamentals of physics remain the same from all perspectives. The alternative—that physical laws should vary with the observer's motion relative to some undetectable, unknowable frame of reference—now seems absurd. Yet every natural philosopher before Einstein, going back to Aristotle and beyond, accepted some version of that proposition.

Einstein arrived at special relativity almost purely from an examination of logical flaws in the then current theories of physics, flaws that were evident for all to see. Speaking recently at the Aspen Center for Physics, physicist Murray Gell-Mann marveled at Einstein's ability to take James Clerk Maxwell's equations of electromagnetism more seriously than Maxwell himself did and track the full implications of Galileo's idea of relative motion and Newton's model of gravity. Here was the glorious payoff of Einstein's fierce commitment to freethinking. He insisted on examining the workings of the world at a more rigorous level than even the most illustrious of his predecessors, until he was totally certain that the system made sense. His requirement of total consistency forced him to take seriously the problems that his predecessors and colleagues alike had swept aside as trivialities or unanswerable matters of metaphysics.

Einstein's triumphs were guided not by strange new experiments but by rigorous logic, not by the most esoteric questions of the day but by the most basic ones. He famously argued that "all physical theories, their mathematical expressions notwithstanding, ought to lend themselves to so simple a description that even a child could understand them." Similarly, he described the questions that motivated his theories as fundamentally childlike ones that he had carried with him into adult life.

Often Einstein framed these questions in terms of thought experiments that highlighted the universal nature of his thinking. In one conceptual exercise, he wondered what a person would see if he could catch up with a beam of light. Newton's theory of space says that such a thing should be possible; Maxwell's theory of light says that it should not. Special relativity showed how to explain what we observe (Maxwell's laws) by discarding what we cannot observe (Newton's absolute space). These mental pictures convinced Einstein that the cosmos had to operate on a more consistent and fundamentally simple basis than his colleagues believed. He addressed these shortcomings by inventing his own physics and finding new ways to measure space and time.

One of Einstein's greatest inspirations, as he later recalled, occurred during a flash of insight in 1907: "For an observer falling freely from the roof of a house there exists—at least in his immediate surroundings—no gravitational field." (A note of caution: This may be the symbolic Einstein at work, retroactively fabricating an event to make his concepts clear. What matters, however, is that his ideas can be expressed in such terms and that he chose to do so.) In other words, the acceleration caused by gravity exactly erases the force exerted by gravity. Einstein cast the situation another way. A man in an enclosed elevator cannot in principle say whether he is motionless on Earth's surface and feeling the pull of gravity or moving through space, being pushed upward at an identical rate of acceleration. This became Einstein's principle of equivalence, which states that a uniform acceleration is equivalent to, or indistinguishable from, a uniform gravitational field.

Einstein developed those ideas into general relativity, a theory of gravity that subsumed and expanded on Newton's. General relativity far surpassed special relativity in redefining space and time—not just how they are measured but how they are linked together—to find a clearer, more comprehensive description of reality. The warping of space-time is an exotic-sounding concept that is actually a part of everyday experience. It is what holds together giant clusters of galaxies, but it is also what I experience every time I sit down in a chair or take a step. The symbolic Einstein showed that physicists need not be remote and detached from the real world; the scientific Einstein showed that their research need not be, either. No wonder he makes so many surprise appearances in my life.

Relativity—general relativity in particular—leads me directly from the scientific Einstein to the philosophical Einstein. With general relativity, Einstein completed a program begun in ancient Greece to determine the scope of natural law and thereby define the relationship between us and the universe. In the Greek conception, the flawed earthly laws and elements are distinct from those of the heavenly spheres, which follow perfectly circular motion and consist of aether—the perfect "fifth element." The perceived split between heaven and earth lived on in diminished form, all the way through Newton's absolute space (which he described as "the sensorium of God") and the nineteenth-century reconception of the aether as an invisible, all-pervasive medium that transmits light through empty space and provides the background reference frame of all motion.

In Einstein's universe there is no fifth element, which means that there is no escaping the authority of science. To me, this is one of the most disruptive aspects of Einstein's entire rebellious vision. Shortly after completing general relativity, he published a paper that expressed these ideas in rigorous terms, essentially inventing the field of cosmology, the study of the cosmos as a whole. He also established the terms for a new relationship between science and religion.

The idea that the universe is a single thing, governed by a single set of rules accessible to mathematics, strikes me as stirring, terrifying, and intensely mystical—a word that undoubtedly would have caused Einstein to wince or laugh, or perhaps both. "Mysticism is in fact the only reproach that people cannot level at my theory," he once retorted when a fan praised him for this aspect of relativity. Yet his snipe against mysticism told only half the story, since the outspokenly atheistic Einstein frequently adopted the language of theology. As in politics, as in science, he reached for a deeper truth by redefining and extending commonly used terms. "What I see in Nature is a grand design that we can understand only imperfectly, one which a responsible person must look at with humility," he said. "This is a genuinely religious feeling and has nothing to do with mysticism."

Once again Einstein plays the role of thoughtful revolutionary, reinventing familiar terms to expose broader truths. He implicitly argued that science (aided in no small part by his theories) had expanded to the point where it redefined not only humanity's relationship to the universe but also humanity's relationship to the divine. Einstein's cosmos leaves no place for a literal heaven, no physical realm where our earthly laws of physics do not apply. But in religion as in science, when Einstein overthrew the old order he exposed a new, deeper order. He found a religious interpretation of this deeper order in the philosophy of Baruch Spinoza and came to regard physical law itself as divine. "I believe in Spinoza's God who reveals himself in the harmony of all that exists, not in a God who concerns himself with the fate and actions of human beings," he said.

Einstein's much repeated use of the word "God" was not an indulgence and not a purely symbolic act. It was a well-considered philosophical position. He acknowledged that a truly universal theory of physics has theological implications; at the same time, he worried intensely about the destructive power of religions whose adherents imagine they can pray for their success or for others' failure. Einstein believed, passionately if a bit naively, that his logical approach could help here, too. "After religious teachers accomplish the refining process indicated, they will surely recognize with joy that true religion has been ennobled and made more profound by scientific knowledge," he wrote in 1941.

I admire the dogged conviction Einstein displayed as he parsed the meaning of God and religion again and again to clarify his self-proclaimed "new religion." Just as his belief in beautiful, orderly scientific theories mirrored a child's view of the world, so his belief in God as the ultimate manifestation of that order expressed an idealistic notion that God is so much greater than humankind that He cannot be found in any one faith. Einstein devoted great energy to publicizing this view. He repeatedly described the "cosmic religious feeling" that accompanies great scientific discoveries and declared in the *New York Times Magazine* that "in this materialistic age of ours the serious scientific workers are the only profoundly religious people." Although there is no deity to communicate with in Einstein's universe, he presented the possibility of a cosmic connection based on an intellectual comprehension of the rules of reality.

So far, this path to spiritual enlightenment is a lonely one. The theoretical legacy of Einstein's foray into cosmology is everywhere. General relativity provided the underpinning of the Big Bang and introduced the concept of a cosmological constant, the model for the "dark energy" thought to be causing the expansion of the universe to accelerate. Modern cosmology depends so thoroughly on Einsteinian notions of curved space-time, the large-scale homogeneity of matter, and the equivalence of all reference frames that many scientists forget that these ideas were radical speculations less than a century ago. Try looking for Einstein's philosophical legacy, however, and the cupboard looks rather bare.

I cannot recall a researcher ever discussing the cosmic religious feeling. Many scientists and historians dismiss Einstein's use of the terms "religious" and "God" as sloppy shorthand for the beauty of science. Cosmologists today rarely talk about God; if they do, it is in the self-conscious manner of Stephen Hawking, who once asked "What place, then, for a creator?" They largely ignore Einstein's philosophical language and the broad, emotive way in which he spoke about his research. In his groundbreaking book *The Inflationary Universe,* for example, Alan Guth of MIT, who codeveloped the leading model of the Big Bang, speculates provocatively about whether it would be possible to create a new universe in a basement laboratory, but he treats such a Genesis-on-demand merely as a scientific thought problem. As cosmology grows ever larger, its aesthetic grows paradoxically smaller and in many ways more impoverished.

This is the saddest aspect of Einstein's legacy. Politicians and activists have taken up his dream of a peaceful, unified world. Physicists have carried on his program to unify the laws of nature. The search for uniformity and harmony, which Einstein regarded as the core aesthetic of his science, guides almost all advanced ideas in physics today, from the most far-out theories of universal beginnings to string theory. The unification of science and religion, by contrast, has drawn few takers. The uncompromising rigor that Gell-Mann lauded in Einstein's science drew little support when Einstein applied it in a theological direction. The modern resurgence of religious extremism seems only to have driven people farther from Einstein's ideal.

Perhaps it is just a matter of time. Smashing icons is never a popular business. Special and general relativity, among the grandest theories of our time (rivaled only by quantum mechanics and Darwinian evolution), took years to gain wide acceptance and never received a Nobel Prize. Cosmic religion is far more controversial and far less concrete: there is no equivalent of a solar eclipse experiment to show that Einstein was on the right track in pursuing "the secrets of the Old One." Meanwhile, I continue to commune with Einstein and do what little I can to follow his uncompromising creation: a science that rejects patchwork theorizing but also rejects the notion that rational inquiry cannot speak to the human hunger for spiritual truth.

In Search of Einstein

LEE SMOLIN

LEE SMOLIN is a founding member and research physicist at the Perimeter Institute for Theoretical Physics, in Waterloo, Ontario. A prominent contributor to the subject of quantum gravity, he is also the author of *The Life of the Cosmos* and *Three Roads to Quantum Gravity*.

In my office hangs an original print of a photograph of Albert Einstein, taken by the Canadian photographer Yousuf Karsh. It was given to me by an esteemed older friend, who had known the photographer, to welcome me to Canada. I have put it in a place of honor, but I rarely look at it. It shows me an old man who gazes at the camera with a mixture of sadness and intensity. You can see in the eyes a fierce intelligence and rebelliousness muted by time—or perhaps just by resignation at having to sit once again in the service of his public. Looking at it now, what I am most sure of is that I have never met anyone like him.

Whether this is because of his singularity or because he was very much of an era that is gone, I don't know. I am also sure that I have no idea who he was, no sense of what he would sound like or how he took up the space around him, what he looked like alive and moving. In 1979, as a newly minted scientist, I was obsessed with learning something about who he was. My first postdoctoral position was at the Institute for Advanced Study, and one of my main reasons for accepting it was the hope of making contact with some living legacy of Einstein. In this I was disappointed. By that time there was no trace of him there, apart from a bust in the library. No student or follower of Einstein could be found. There were a few people who had known him, but they discouraged my attempts to question them.

My first week there, the theoretical physicist Freeman Dyson came by and invited me to lunch. Very much the gentleman, he asked if there was anything he could do to make me more at home in Princeton. I said I had but one request: "Could you tell me what Einstein was really like?"

"I'm very sorry, but that's one thing I can't help you with," Dyson replied.

Surprised, I insisted: "But you came here first on a fellowship in 1948, and you were a colleague of his until he died in 1955."

Dyson explained that he, too, had arrived at the Institute with a strong desire to know Einstein, so he went to Einstein's secretary, Helen Dukas, and made an appointment. The day before he was to meet Einstein, he thought he ought to have something to discuss, so he asked for and got from Miss Dukas copies of Einstein's recent scientific papers. Reading them that evening, he decided they were junk. (This was indeed the way Einstein's work on a unified field theory was then regarded by most of the younger generation; how ironic that now the Institute is filled with young people playing with unified field theories.) The next morning he decided he couldn't face Einstein and tell him his work was junk, but he couldn't face him and *not* tell him, either. So, he told me, he skipped the appointment and spent the next seven years until Einstein's death avoiding him.

Stunned by this story, I could only say the obvious: "Don't you think Einstein could have defended himself and explained his motivation to you?"

"Certainly," Dyson replied. "But I was much older before that thought occurred to me." He went on to describe the arrogance that afflicted the members of his generation of physicists once they had made quantum field theory work, and how easily they dismissed anyone who did not agree with them.

Shortly after that, a friend introduced me to Abraham Pais, who was coming regularly to the Institute to consult Einstein's papers for his biography. Pais launched into a soliloquy in praise of older women, and when I asked whom he was referring to, he replied, "Of course, Miss Dukas. They don't make women like her anymore." After a long acquaintance, Miss Dukas had just allowed Pais to address her with the familiar *du,* and this thrilled him. She, it turned out, was still at the Institute, coming in every day to work on the organization of the papers. I asked Pais if he would introduce me so I could see for myself what he meant, and the next day at lunch he obliged. Left alone with her, I told her of my desire, as a young physicist trying to follow Einstein's path, to have some sense of what he was like as a person. Her expression turned icy. "I can't help you," she said. "I was just the secretary. I only typed the papers." I might have argued that having lived in the same house with him for more than two decades, she certainly could tell me what I wanted to know, but our interview was over. The next year, she died.

I got nowhere as well with Peter Bergmann, one of Einstein's former assistants and my colleague later on. To the same question, he too insisted that he couldn't help and changed the subject. Nor was Pais any more forthcoming, although he was otherwise friendly to a young scientist who shared his love of contemporary art. John Wheeler told me some wonderful stories of his encounters with Niels Bohr and Einstein, but somehow no picture emerged of Einstein the human being.

So I never found out who Einstein was. Perhaps there was a deliberate attempt to suppress impressions that would interfere with the myth that his heirs wanted to last for ages. Perhaps they were oversensitive, given how roundly Einstein had been scorned by the physics community in his last years at Princeton. Or perhaps those who had been through two horrible wars, which had destroyed utterly the world they once lived in, found it impossible to communicate with a young American who could have had no conception of what it felt like to walk the streets of Berlin on a spring day in 1925.

A few times, finding myself alone in the library late at night, I gazed at his bust. It told me as little as the abstract metal sculpture down on the lawn. Both commemorated a myth; there was no opening, no clue as to the character of the man himself.

Perhaps it was arrogant to want some sense of the person Einstein was, but I felt I had some reason to ask, for my decision to do theoretical physics was entirely due to a single encounter with Einstein's writings. One day, as a seventeen-year-old high school dropout in Cincinnati, I was led by my interest in architecture to a book of essays on Einstein called *Albert Einstein, Philosopher-Scientist*, edited by Paul Schilpp. (The path had led from Buckminster Fuller to geodesic domes to the mathematics used to construct them and thence to the mathematics of Einstein's general theory of relativity.) The book is prefaced by a brief autobiographical note by Einstein, in which he describes his motivation to do science, which was apparently based on an adolescent angst and rebellion not unlike my own:

> When I was a fairly precocious young man I became thoroughly impressed with the futility of the hopes and strivings that chase most men relentlessly through life. . . . By the mere existence of his stomach everyone was condemned to participate in that chase. The stomach might well be satisfied by such participation, but not man insofar as he is a thinking and feeling being.

He goes on to describe the answer he found:

> Out yonder there was this huge world, which exists independently of us human beings and which stands before us like a great, eternal riddle, at least

partially accessible to our inspection and thinking. The contemplation of this world beckoned as a liberation, and I soon noticed that many a man whom I had learned to esteem, and to admire, had found inner freedom and security in its pursuit. The mental grasp of this extrapersonal world within the frame of our capabilities presented itself to my mind, half consciously, half unconsciously, as a supreme goal. Similarly motivated people of the present and the past, as well as the insights they had achieved, were the friends who could not be lost.

I resolved then and there to follow his path and become a theoretical physicist. Although until that moment I had never considered doing any kind of science, that evening—with the reckless clarity of adolescence, which makes possible all kinds of improbable projects, from war to rock 'n' roll—I decided I would dedicate my life to continuing his search for the laws of nature. Although I had never taken a physics course of any kind, somehow I felt this was something I could do. I took on his main unfinished project: to find a more rationally comprehensible substitute for quantum theory and to do this by understanding quantum physics as a consequence of the structure of space and time. I can only marvel at how lucky I was to have encountered Einstein at this point, for this search has been the basis of my scientific work ever since.

I started right away to teach myself physics, with Einstein as my tutor. I found a book containing the original papers on relativity and read it straight through. I have a very strong memory of reading Einstein's 1907 paper on the equivalence principle on the subway and the shock of understanding his argument for light bending. Meanwhile I got a catalog from MIT, from which I was able to make an outline for how I would teach myself physics and then apply to graduate school. Luckily a friend of the family, a mathematician named William Larkin, convinced me that I would do better first to go to college. I had actually already been admitted to Hampshire College, in Amherst, Massachusetts, on the strength of my interest in architecture. I stopped there on my way to MIT and met the new physics professor, Herbert Bernstein, who spent an afternoon pretending to know nothing about general relativity and letting me talk about it. That conversation, and a letter he sent me afterward, convinced me that I had found a living teacher—and, indeed, without him I would certainly not have learned how to do the hard work physics research requires. Later that spring I took a graduate course in general relativity from Paul Esposito, which was the start of my real education in physics. There are not many physicists who learned general relativity first and then understood Newtonian mechanics as an approximation to it, but future curriculum planners may be interested to know that it works.

Professor Bernstein had already instituted a revolutionary curriculum, which was to teach quantum mechanics as freshman physics and then, in a later course, derive classical mechanics from it. I had learned of quantum physics from Einstein's critique of it in his "Autobiographical Notes," and I had followed that by reading in the same book an article by Bohr on his debates with Einstein, so I knew the controversies over interpretation before I knew the details. I was eager to learn the real thing, and it also helped that, unlike many professors, Bernstein did not shrink from including the controversies in his course. At the end of it, we read the key 1935 paper by Einstein, Podolsky, and Rosen, in which they argue for the incompleteness of quantum mechanics. This is also the paper that discusses the phenomenon of two particles' being irreducibly entangled once they have interacted. I remember reading the paper on the college lawn one spring afternoon and being deeply struck by the implication that my atoms were entangled with those of every person I had ever touched.

I reread that paper many times and also Bohr's reply. There was a corner of my room where the ceiling met the walls at an odd angle, which I used to stare at while contemplating particle entanglement. I also found in the library the report of the Solvay Conference of 1927, with transcripts of the debates between Einstein and Bohr and their discussions with their colleagues on the quantum theory, and I read every word carefully. I found Bohr's reasoning fascinating but in the end unconvincing. Einstein by that time had persuaded me that quantum mechanics is incomplete and requires replacement by a new theory, and this is still my view.

There was never a question in my mind that Einstein was right. Although I respect my colleagues who disagree, I find their thinking basically incomprehensible. As much as I try to see what they are talking about, I find the assertion that nature is actually a vector in a complex space made up of infinite dimensions as silly as Aristotle's universe of concentric spheres surrounded by heaven with Earth at the center. Is the fact that measurements return definite values just a consequence of our manipulations? Did the universe wait almost 14 billion years for the descendants of the ape to decide to do experiments before its wave function collapsed? Is the world just information waiting to be decoded? I have worked with quantum mechanics all my life and it still makes as little sense to me as it did the first year I learned it. So I take some small comfort in the fact that it never made sense to Einstein, either.

From that point on, Einstein became a kind of imaginary mentor. His demand for clarity above all else, and his insistence on pursuing a path toward truth based on his own judgment and without regard to what people around him were doing or thinking, served as my role model. His fierce independence also encouraged me to think that there was at least a chance to succeed in science by finding your own way.

At the same time, I should admit that my own career has been possible because I did *not* follow Einstein. Had I approached physics with the same moral purity he did and refused to work on quantum physics because I found it unacceptable—had I insisted on working only on ideas of how to invent a theory to replace it—it is likely that I would have suffered the same fate he did, which was to have no academic position after graduation. Instead, I took a detour that I am only now coming out of. I decided there was little to be gained either scientifically or professionally from a head-on assault on quantum theory; instead, I decided to attack the problem of combining quantum theory with gravity. As gravity was the only physical force that had not been successfully incorporated into quantum mechanics, I felt that clues to a replacement theory would most likely be found in the attempt to extend quantum mechanics to gravity and general relativity. This was certainly a path Einstein would have laughed at, and I don't think that the partial success we have had would have changed his mind.

But while I was departing from Einstein's path, I kept encountering him. In graduate school at Harvard, I met a woman who studied the history and philosophy of physics. We dated by reading together Einstein's early papers in the order in which he wrote them. For reasons known only to itself, the Einstein estate has blocked the publication of a book that would do the most to spread Einstein's influence among students—an inexpensive paperback containing all of Einstein's key scientific papers translated into English. But with my friend's German we were able to read them, and this supplemented the education I was getting in my courses on quantum field theory.

From those papers I absorbed something of Einstein's scientific style. One thing rarely appreciated is that the young Einstein, unable to get a job in his profession, was actually very good at the craft of theoretical physics. The math is for the most part simple, and the physical argument is confident and precise. The structure is spare: there is nothing extra, no repetition or piling on of examples or evidence. The introductions are brief and to the point;

experimental results are mentioned but not dwelt on. Each paper presents a simple but devastatingly effective argument, often with a simple calculation leading to a surprising but robust and far-reaching conclusion. Reading these papers is a little like looking at a drawing by Picasso, watching a play by Beckett or Brecht, or listening to Thelonious Monk play the piano. The power is in what is held back, what is not mentioned, so that the few lines or notes necessary to carry the thought can do their work on us.

This style is the most humbling for those who understand what is going on, because while one could easily use more sophisticated math, or calculate more, or display more control over the details, the only way one could emulate Einstein is to think better—and that is not a matter of education. Thinking well is much harder than anything else we do, which is why many of us will do almost anything to avoid having actually to think about what we are doing and why. We need to face the uncomfortable truth that one person, having failed in his search for an academic position, in the spare time he could steal from his job and family, taking walks and sitting in his room alone, did far more in five years to advance science than a thousand string theorists have done in the past twenty years, secure in the best conditions in the world's best universities and institutes.

That is the Einstein I would really like to have known—not the old sage (or whoever he was) sadly offering his visage for yet another photographer or sculptor, but the young man with a confused love life who had failed to impress his professors but was proceeding confidently to remake science. We have only a few pictures of him stuffed into a suit at his desk in the Bern patent office. In them, the mouth is sad but the eyes are dangerous, even uncivilized. Looking at them, I can only think that I have no idea who he is or whether I would even like him. I know enough to know I am nothing like him, even if he gave me a small part of his moral quest for knowledge to serve as my life's purpose.

Living in a luckier time, I suspect I am a happier person than he was. But were he, impossibly, to come to life, I know I would care far more for what he thought about the little my colleagues and I have been able to contribute to his unfinished work than for any form of recognition my profession could offer.

Einstein and Absolute Reality

ANTON ZEILINGER

> ANTON ZEILINGER is a professor of experimental physics at the University of Vienna whose research interests include photonic entanglement, matter-wave interferometry, and quantum information theory. He is also coeditor of *The Physics of Quantum Information Theory: Quantum Cryptography, Quantum Teleportation, Quantum Computation* and *Quantum (Un)speakables: From Bell to Quantum Information.*

Like many others, I was fascinated by Einstein at a very young age. But when I heard in high school of the Michelson-Morley experiment disproving the existence of the aether, the presumed medium for the transmission of light, and learned how Einstein had solved the mystery of its absence by developing his special theory of relativity, my life was changed. I decided to become a physicist. I still remember my excitement when, on a Sunday afternoon hike with my parents and my sister in the Lainzer Tiergarten in Vienna, a game park that used to be a hunting ground for the Hapsburgs, I suddenly understood how the argument went. It struck me out of the blue! How in the world could the universe be so strange that the speed of light was the same no matter how fast I, the observer, was moving or how fast the light source was moving?

But what was really striking was the idea behind the argument: accept as a physical concept only that which is measurable, that which is observable. Thus, time is what is measured by clocks—and there is no way to tell, while you are moving, whether your clock is going faster or slower than a clock at rest. Therefore there is no reason that clocks in different states of motion relative to each other should keep the same time, such a difference actually being predicted by Einstein's special theory.

Years later, I learned that Einstein's viewpoint can be seen as a special case of a general principle proposed by the Austrian philosopher-physicist Ernst Mach: Any statement in physics must be a statement about relations between observable quantities. Thus, just as in consulting a clock one cannot say whether it runs slow or fast on an absolute scale, absolute time, as postulated in the seventeeth century by Isaac Newton, does not exist. One can only make statements about how fast different clocks go relative to one another, depending on whether they are moving or at rest with respect to one another. The concept of absolute time thus does not make sense unless there is a way to observe absolute motion. As the latter is not possible, the first cannot exist.

Mach's principle appeared to me to be essential for all of physics. Physics—and all of science, for that matter—should never make statements about something that cannot be observed even indirectly, and it should make statements only about how different observations relate to one another. Long after my student days, it came to me as a total surprise that Einstein had not applied this (in my eyes, his own) fundamental approach, which was so successful in his work on relativity theory, to quantum physics as well. Instead of accepting only concepts that can be verified by observation, Einstein insisted on the existence of a reality prior to and independent of observation.

I learned quantum physics very late. I did not spend a single hour in a class covering quantum physics or quantum theory. At that time (the 1960s) the physics curriculum at the University of Vienna was very free. There was no bachelor's or master's degree; you went directly for the Ph.D. At the end, there was a lengthy exam requiring you to know physics; it did not matter where you got your knowledge from. So for that exam I learned quantum physics, but only from books. Whether or not that was an advantage I don't know. But I immediately fell in love with quantum physics because of its immense mathematical beauty. On the other hand, I also felt that something was missing. There was very little discussion in those books of the fundamental philosophical issues raised by quantum theory.

This changed years later—again, when I ran into Einstein's work—in 1976, at a conference held in the beautiful Sicilian village of Erice. The meeting, called "Thinkshop on Physics," was organized by John Stewart Bell and Bernard d'Espagnat, and it was devoted to the foundations of quantum physics. There I heard for the first time of the Einstein-Podolsky-Rosen (EPR) paradox. There was talk of reality, of hidden variables, of entanglement. There were intense philosophical fights, and I did not understand anything, and my head started to spin. All this was too new to me. But my interest was kindled.

When I finally read the EPR paper of 1935—the paper in which entanglement is introduced—I understood very little. I read it again and again. Today I know that there were at least two problems with it. One is that the paper is written in a rather complicated style. Einstein himself, in a letter written the same year to Erwin Schrödinger, regretted that its main point remained hidden behind too much *Gelehrsamkeit* (erudition). The other problem, which probably any modern physicist encounters in reading the paper, is that one has to shed all one's quantum instincts in order to get through it. If one has already become comfortable with the way quantum physics works, it requires a strong effort to understand the EPR reasoning. In any case, I slowly began to appreciate the challenge that entanglement poses to our cherished notions of reality.

Einstein's views on quantum physics are often misunderstood. He must be credited for his enormously deep insight, even though he eventually turned out to be wrong. He must be credited first because he pointed out important concepts regarding the way in which quantum physics clashes with our ingrained views of how the world operates. And he must be credited because the issues he raised gave rise to many experiments that are laying the basis for a new information technology that entails concepts such as quantum cryptography, quantum teleportation, and the quantum computer.

Though I had known for a long time that Einstein was a realist—some would even call him a stubborn realist—I discovered only recently how deeply realistic Einstein's view of photons was from early in his career, as expressed in a very explicit way in his famous paper of 1905 entitled "Über einen die Erzeugung und Verwandlung des Lichtes betreffenden heuristischen Gesichtspunkt" ("On a Heuristic Point of View Concerning the Production and Transformation of Light"). That aspect of the paper is not widely known. The paper is generally seen as the starting point of the photon idea, because in it Einstein talks of light quanta. (The name "photon" was invented much later, in 1926, by the American chemist Gilbert N. Lewis.)

Concerning this 1905 paper, a widespread misconception needs to be corrected. The usual view is that Einstein arrived at the photon idea via an analysis of the photoelectric effect. But his approach was a much more subtle one. What he does is compare the entropy of a gas confined to some volume with the entropy of radiation inside a cavity as derived by Max Planck five years earlier (entropy being a measure of the disorder of a system). Now, the entropy of the ideal gas can easily be understood. Following Ludwig Boltzmann, the entropy is given by the probabilities of finding the gas particles within a certain volume; the more confined the gas is, the smaller the volume the atoms have to fill, and the lower the entropy. Einstein then observes that in the case of radiation within a cavity, the entropy varies in the same way with volume. Therefore Einstein assumes by analogy that radiation within a cavity must also consist of particles—the quanta of light, as he called them. Only then, using this idea, does Einstein analyze the photoelectric effect and find agreement with experimental observation.

He assumes an extreme realist position when he says that "the energy of a light ray spreading out from a point source is not continuously distributed over an increasing space but consists of a finite number of energy quanta which are localized at points in space, which move without dividing." Today we know that the idea of anything localized in space is a problematic idea, and for light we know that it is a theoretical impossibility. We have learned that in quantum physics we should attribute only those features to a system which have actually been observed in experiment.

That the Nobel committee awarded Einstein his Nobel Prize for the 1905 photon paper and not for the theory of relativity is often thought to be shortsighted. But evidently Einstein himself was aware of the special character of this paper. In a letter to his friend Conrad Habicht he writes that it "deals with radiation and the energy properties of light and is very revolutionary." In that letter he also mentions his other annus mirabilis papers—the one in which he proposes his special theory of relativity and the papers about atoms and Brownian motion—but these he does not call revolutionary. So it seems that the Nobel committee gave him the prize for what he himself considered the most revolutionary.

Einstein's criticism of quantum theory is often presented as if it were due to some lack of understanding. But it is the opposite. In my opinion, his criticism arose from his very clear realization of the enormous implications of the new theory for the world of physics. Einstein started to worry very early about the implications of the quantum idea. I find it impressive that as early as 1909 he expressed his *Unbehagen* (discomfort) about the role that randomness, or chance, plays in quantum physics—a role going significantly beyond that in classical physics (and in daily life, for that matter). This is all the more remarkable because it took until 1925–26 for the full quantum theory to be finally worked out by Werner Heisenberg and Erwin Schrödinger.

The discovery that (trivial exceptions aside) quantum physics makes only probabilistic predictions is certainly one of the deepest philosophical discoveries of science. After all, the program of science over the centuries has been *investigatio causarum,* the investigation of causes. And after centuries of digging deeper and deeper along the causal chain, we finally came to a stop. The individual quantum event happens by chance. There is no hidden cause, no hidden reason. But fundamental randomness is unbearable to us. Whenever something, anything, happens, we always ask for the reason—why it happened just this way and not that way. We don't give up until we find comfort in settling on some cause, no matter how implausible it might be. And now, suddenly, quantum physics tells us of events that just happen—happen without any specific cause. Einstein was disturbed by this. He supposedly once exclaimed that if that randomness remained with us, he would sooner work in a casino than as a physicist. It is therefore rather amusing that, based on the randomness of the path a photon takes after a beam splitter, we can today build random-number generators that produce sequences of random numbers more random than those produced by any algorithm or by any classical physical device. So Einstein's photon may well be put to work in a casino someday.

I mentioned that Einstein himself had reservations about the way the EPR paper was written. For me the issue of entanglement is much better explained in Einstein's "Autobiographical Notes," published in 1949 in P. A. Schilpp's beautiful collection of essays entitled *Albert Einstein, Philosopher-Scientist.* Starting from two particles that interacted at some time in the past, there is a set of measurements in which the results, for both particles, are perfectly correlated. So, for example, if we measure the position of particle 1, we know exactly the position of particle 2, and if we measure the momentum of particle 1, we know exactly the momentum of particle 2. Quantum-mechanically, we therefore have to assign particle 2 a different quantum state after the measurement of particle 1. In the first case, it is a state with definite position, in the second case a state with definite momentum. Now, and this is Einstein's point, "the real factual situation of the system S2 is independent of what is done with the system S1, which is spatially separated from the former." He then argues that since we assign two different quantum states to system S2 depending on the specific measurement performed on system S1, the quantum state cannot be a complete description of the "real factual situation."

This kind of reasoning is impeccable as long as we assume, as Einstein did, that it makes sense to consider a "real factual situation" per se—that is, independent of observation. But do we have to make that assumption? In my opinion such a position cannot be justified in a non-operational way; that is, reality cannot be defined without recourse to experimental observation.

Most interesting, in investigating experimentally the foundations of quantum physics, the basis was laid for new concepts in information technology. For example, the combination of objective randomness with entanglement is at the heart of two modern concepts: quantum cryptography of the Ekert protocol type and quantum teleportation. In entanglement-based quantum cryptography, the two players (conventionally known as Alice and Bob) share entangled pairs of particles (usually photons). As soon as Alice measures her photon, one of the two orthogonal polarizations, say, results, and Bob's photon instantly assumes the identical polarization. Entanglement guarantees that Alice and Bob obtain the same random result. Having shared many pairs, they obtain a long sequence of random numbers, which they then can use as the key to encode a secret message. (Besides providing the random sequence, the randomness of the individual event also guarantees that any eavesdropper can be caught.)

In teleportation, we have a double application of entanglement. First, Alice and Bob, anticipating their wish to teleport a photon, share an auxiliary pair of entangled photons. Then Alice entangles the original photon, the one whose state she wishes to teleport, with her member of the auxiliary pair. This entangling procedure instantly transfers the information carried by Alice's original photon over to Bob's member of the auxiliary pair. The procedure suggests that information has been carried faster than the speed of light, thus violating Einstein's relativity theory. Fortunately, quantum randomness comes to the rescue. Alice can never, not even in principle, build an apparatus able to force her two photons into a *specific* entangled state; rather, there are four possible entangled states, and she has no influence over which one will result. Once again, this is complete quantum randomness. Depending on the specific entangled state obtained by Alice, Bob has to rotate his particle in a specific way in order to obtain the original photon. And as long as he does not have Alice's results—which cannot arrive faster than the speed of light—he cannot read any information out of his photon.

It would be interesting to know Einstein's reaction to the state of quantum physics today. While he would certainly be pleased that his photons have been put to work in a way no one of his time could have imagined, the philosophical and conceptual problems have become even more pressing. We know today that individual events, at least those on the micro scale, really are random, and the phenomenon of quantum entanglement tells us that quantum mechanics is not completable in the way Einstein would have liked.

Nowadays, with the idea of a reality that exists prior to and independent of observation losing more and more ground, it might be productive to leave such a notion behind us for good. After all, there is no way to learn anything about reality without observation, and the concept of an unobservable reality permanently beyond our reach is a concept devoid of any meaning. Do we have any hint of a reality existing independent of observation? I suggest that the strongest signature of such a reality would be something beyond any possibility of being influenced by us but which can nevertheless be observed. We note that such a thing actually exists. It is the outcome of an individual quantum measurement, which, since it is completely random, is beyond our influence. Thus the very randomness Einstein disliked so much turns out to be a strong indication of the independent reality so prized by him. But to talk about reality without explicit reference to the act of observation is devoid of meaning. Einstein would certainly have disliked such a strange concept of reality.

A Walk down Mercer Street

STEVEN STROGATZ

STEVEN STROGATZ is a professor in the Center for Applied Mathematics and the Department of Theoretical and Applied Mechanics at Cornell University. He did his doctoral work in applied mathematics at Harvard and stayed for three years as a National Science Foundation postdoctoral fellow. From 1989 to 1994 (when he went to Cornell), Strogatz taught in the Department of Mathematics at MIT. He has received awards for both his teaching and his research, including MIT's highest teaching prize, the E. M. Baker Award for Excellence in Undergraduate Teaching, and a Presidential Young Investigator Award from the National Science Foundation. He is the author of *Sync: The Emerging Science of Spontaneous Order* and the textbook *Nonlinear Dynamics and Chaos*.

I can still remember—or at least I believe I can remember—when I first became aware of Albert Einstein. It was in second grade at North School in Torrington, Connecticut. Every week, our teacher, Mrs. Crowston, would march us all to the school library and tell us to choose a book. And every week I'd pull the same one off the shelf: *The How and Why Book of Atomic Energy*.

We were not allowed to take the books out of the library, but with constant rereading I managed to memorize my favorite parts of it. Somewhere in the old folder my mother kept of my school assignments from those years, there's a flimsy scrap of yellowing paper with blue-lined rules on it, one of the first essays of my life.

I've just gone looking for that paper and found it, packed deep inside a file in a box in the attic. An annotation is tucked in the upper left-hand corner, penciled in maternal cursive script: "Steven. 7 yrs. old. April 1967." And then, in my beginner's printing, starting on the top blue line, in large, dark letters: "The Atom Bomb." A mushroom cloud rises from the *m* in "Bomb." The letters *B, o,* and *m* are written in scary, squiggly, horror-movie-poster style, but for some reason I forgot to squiggle the final *b.* Then, in lighter, smaller print, with one sentence per line:

> Einstein invented the Atom Bomb.
> One Atom Bomb could blow up half of Rhode
> Island!
> Einstein's theory was $E = mc^2$

Finally come four empty lines and a parenthetical caption—"(Picture of the atom bomb)"—with an arrow pointing to a drawing that looks like a torpedo, except that it has a bright red cap penciled on it.

The back side of the paper, which I'd forgotten about completely, seems to be an attempt to write a book on one page. It's headed "Science of Atomic Energy" and includes factoids—"Atom Smashers split atoms so fast the limit is about 300,000 of a wink of an eye! (5,000 1/2 of a second)"—and a list of radioactive elements, apparently recalled by heart, because some are misspelled: "Plutonium neptunium Radium Uranium cobalt sturtbium truium."

My parents' friends found all this hilarious. "What do you want to be when you grow up?" they'd ask. "A nuclear physicist," I'd say. And invariably came the reply: "Gee, I can't even *say* that!"

So at that point Einstein was a weapon for me—a way to show off. I understood that everyone thought he was smart, maybe the smartest person of all, and that by wanting to be like him I'd seem smart, too. Is that too cynical? Am I being unfair to my seven-year-old self? Maybe. Because I did genuinely love science, and devoured the encyclopedia and books about lightning or dinosaurs. Einstein symbolized all that for me.

There's an old Jesuit saying: "Give me the child until he is seven, and I will show you the man." In looking back on my life's trajectory, I'm indeed struck by how little has changed. Just seven years after that initial essay, at fourteen, I became obsessed with Einstein, fascinated to the point of hero worship. I don't remember exactly why, but I do remember feeling a sort of kinship with him, especially with regard to his sense of awe about the universe.

Near the beginning of Banesh Hoffman's biography, *Albert Einstein: Creator and Rebel*, which was my bible in those years, he quotes Einstein telling a story that he apparently told often, about a mystical encounter with a compass needle as a child of four or five. Hoffman writes:

In his autobiography the aging Einstein vividly recalled the sense of wonder that had overwhelmed

him those many years before: here was a needle, iso-
lated and unreachable, totally enclosed, yet caught
in the grip of an invisible urge that made it strive
determinedly toward the north. Never mind that
the magnetic needle was no more wonderful—no
less wonderful—than a pendulum striving toward
the earth. . . . It did not fit. It mocked his early,
simple picture of an orderly physical world. In his
autobiography he wrote, "I can still remember—or
at least I believe I can remember—that this experience made a deep and abiding impression on me."

Reading those words today, it's hard to remember what
I must have felt as a fourteen-year-old. The mention of the
pendulum alone must have gripped me. Just the year
before, I'd had my own epiphany about the harmony of
the universe, when I saw that I could connect a mathematical law with the motion of a pendulum. It was thrilling to
imagine that I had felt what Einstein felt. The shared experience of the sublime, the sudden glimpse of an invisible
world—all that must have made me feel close to him.

Or did it? Is memory reliable? Look at the way that
Einstein himself so openly struggles with the mystery of
memory in his quote. If the experience made such a "deep
and abiding" impression on him, Hoffman asks, why was
Einstein so tentative ("or at least I believe I can remember")? Perhaps because he always sought the truth in all
things, and that paradoxical phrasing recognizes an essential truth about memory—its impossible combination of
vividness and fragility. That same tension afflicts me now,

as I try to remember things long past. That's why I chose to borrow Einstein's phrasing in the first sentence of this essay.

Over the next few years of high school, I read everything I could about physics and math, from popular expositions of advanced concepts like group theory and relativity to the history and philosophy of science. I'd curl up in a big soft chair in my high school library, with its burnt-orange carpeting and burnt-orange upholstery, captivated by a compendium of essays called *The World of Mathematics*, a four-volume set of reprinted articles by geniuses such as Poincaré, Newton, and Bertrand Russell. I was in heaven. Meanwhile I kept reading about Einstein. I liked his simplicity and his determination to think for himself, to take on the giants who had preceded him. I especially admired his cockiness. When asked how he would have felt if Arthur Eddington's eclipse observations had not confirmed his prediction, based on general relativity, that starlight would be bent by gravity as it passed by the sun, Einstein is said to have replied, "I would have been sorry for the dear Lord; the theory is correct."

Admittedly, at this stage in my life I wasn't able to understand Einstein's scientific ideas in any depth, but perhaps that didn't matter so much. What I cared about more were the other lessons he taught—about how to act as a scientist, how to feel about God and authority and the wonder of the universe, how to fight, how to be stubborn, how to trust your instincts, and how to admit when you're wrong.

My favorite story was about his epic battle with Niels

Bohr at the 1930 Solvay Congress, where Einstein concocted a thought experiment that seemed to refute Werner Heisenberg's uncertainty principle. Everyone, including Heisenberg, was sweating about it. And then the heroic rescue by Bohr, pondering the paradox all night long, striding in the next day with the answer—Einstein had overlooked something. He'd forgotten to apply one of his own ideas, the weird notion that a clock runs at a different rate when it moves through a gravitational field. When this effect was properly accounted for, Heisenberg's principle was saved. Einstein had been hoist with his own petard.

Einstein's fallibility did not diminish him in my eyes; if anything, it made him seem more approachable—more lovable, even. When he wrote to a student, "Do not be concerned about your difficulties in mathematics; I can assure you that mine are still greater," he wasn't being falsely modest. Of course he was a towering genius at mathematics, but there were plenty of things that even he didn't know and that he found difficult to learn and understand. All of this was reassuring to teenage insecurities.

As fun as it was to read these sorts of stories about Einstein, I also found them a little unsatisfying. I wanted to understand his actual ideas, to follow the train of his mathematical logic, step by step. But there was so much math to learn.

One night, I mentioned to my science teacher, Mr. diCurcio, that I'd read somewhere that when Einstein was a young student he'd been dazzled by something called Maxwell's equations for electricity and magnetism, and I said I couldn't wait until I knew enough math to understand what they were. This being a boarding school, we happened to be at a family-style dinner, sitting around a big table with several other students, Mr. diCurcio's two daughters, and his wife, and he was serving mashed potatoes. As soon as I said I would love to see Maxwell's equations sometime, he put down the mashed potatoes and said, with a mischievous smile, "Would you like to see them right now?" And before I could answer, he grabbed a nearby paper napkin and began writing cryptic symbols on it—dots and crosses, upside-down triangles, E's and B's—and soon he seemed to be speaking in tongues: "The curl of a curl is grad div minus del squared . . . and from this we can get the wave equation . . . and now we can explain what light is." Awestruck, both by the equations themselves and by his command of them, I looked at Mr. diCurcio with new eyes. Until then he had been my teacher, my tennis coach, my friend and mentor. But that night at dinner he revealed another side of himself. Mr. diCurcio knew Maxwell's equations off the top of his head. I was sure that there was no limit to what I could learn from this man. And as it turned out, with his help I learned enough math and physics over the next two years to be on the brink of understanding Maxwell's equations.

Then it was off to college—Princeton, where Einstein had spent the last twenty years of his life. Within days of arriving, I organized a pilgrimage to Einstein's house, at 112 Mercer Street. I had persuaded a few other freshmen to come with me, and we wandered the tree-lined backstreets of residential Princeton, eventually finding ourselves in front of a modest white house with black shutters and a small porch. With heart palpitating, I walked up the steps of the porch and rang the doorbell. An unsmiling old woman answered. I sputtered a few words about how much I admired Professor Einstein.

"Zis is a private residence," she replied, with noticeably guttural *r*'s.

"May I come inside, just for a second?" I asked.

"Zis is a private residence." And gently but firmly she closed the door in our faces. Later I realized who this must have been—Helen Dukas, Einstein's longtime assistant and secretary.

Despite this setback, it was a heady experience being in Einstein's old neighborhood. And now, more important, I was finally prepared to come to grips with his science, even if only in its most elementary form. A moment of truth came in the spring semester, in the introductory course on electricity and magnetism, the course where Maxwell's equations are introduced. The book we used, *Electricity and Magnetism,* by the Harvard physicist Edward M. Purcell, delighted me. It was elegant. Sophisticated. Fun. And best of all, instead of just shoving facts at you, it taught you to think like a physicist. For example, Purcell mentions somewhere that the repulsive force of one electron on another must necessarily lie along the line between them. I'd heard that fact before and always taken it as obvious, but Purcell explained it in a profound way. He said it followed from the isotropy of space—the fact that space has no inherently preferred direction. The argument was that if you imagine two electrons all alone in the universe, only one direction would be singled out—the line between them—and so the force must lie along that line. In other words, the symmetry of space itself was built deep into the laws of electrical repulsion.

This wasn't the moment of truth, however. That would come later, when Purcell considered the problem of electrical current circulating around a long loop of wire. At that point in the course, we had learned about electricity but not about magnetism. Purcell gives a gorgeous argument in which he views the moving electrons relative to two different reference frames. Invoking the Lorentz contraction (one of the central consequences of Einstein's special theory of relativity, which we had already learned about in the fall semester), Purcell shows that Ampère's law of magnetism follows from the laws of electricity, when viewed in a suitable reference frame. In plainer words, electricity and magnetism are not two separate things; they are two aspects of the same thing. And only when you know relativity theory can you see that marvelous unity.

Later I came to realize that Purcell's insight was actually Einstein's. Relativity theory was born of Einstein's desire to reconcile the laws of electricity and magnetism with the laws of motion. Doing so required a total reformulation of our ideas about space and time—now codified in the special theory of relativity.

That was about as close as Einstein and I have ever come scientifically. I never did become a nuclear physicist, or even any kind of physicist at all, though most of my scientific papers today are published in physics journals. My change in direction came on the day I suddenly understood the significance of DNA's double-helical structure. That set me off on a lifelong quest to apply mathematical ideas in biology, in social science, and in other fields where complex systems arise. Nevertheless, I can't quite escape from Einstein's gravitational pull. When I learned about the erratic fluctuations that scientists call noise, there he was, with his early contributions to the theory of Brownian motion and stochastic calculus. And in my studies of how enormous systems spontaneously synchronize themselves, there he was again, with his ideas about stimulated emission (which ultimately led to lasers) or about the strange statistics of particles called bosons (which underpin the theory of superfluids, superconductors, and Bose-Einstein condensates). The more I learn about nature, the more I see of him.

Things and Thoughts

PETER GALISON

PETER GALISON is Mallinckrodt Professor of the History of Science and of Physics at Harvard University. His work explores the complex interaction among the three principal subcultures of physics—experimentation, instrumentation, and theory. In 1997 he was named a John D. and Catherine T. MacArthur Foundation Fellow; in 1999 he was a winner of the Max Planck Prize of the Max Planck Gesellschaft and Humboldt Stiftung. He is the author of *How Experiments End; Image and Logic;* and *Einstein's Clocks, Poincaré's Maps.*

My first idea of Einstein—in fact of anything at all about science—came from my great-grandfather, Frank Alexander, who studied at the Technical University in Berlin and arrived in the United States from Germany late in the nineteenth century. Frank Alexander came from a long line of engineers, but he was the first to switch from civil engineering to electrical and radio engineering. Around the turn of the century, he worked in Thomas Edison's laboratory complex in New Jersey. Then, with various patents he had taken, he launched a small electrical firm of his own in Manhattan, which manufactured various equipment ranging from electric lights to high-voltage testing devices.

Some of my earliest, happiest memories are of going down into his basement laboratory. From a house full of overstuffed chairs and oil paintings, we'd descend into a large, somewhat dimly lit room that had been divided into aisles of long metal floor-to-ceiling shelves stuffed with ammeters and voltmeters, switches, glassware, motors, and coils. For almost a decade—he died when I was fourteen or so—that dusty, hidden redoubt was exactly my image of what a real scientific laboratory should be. Bottles of mercury lined the shelves (I shudder now to think of the many hours I spent playing with the stuff). There were lathes, which spewed out curled metal shavings when he made custom screws and joints for his work. On one wall, like something out of Dr. Frankenstein's work space, were mounted enormous double-throw switches. There were neon and phosphorescent lights he had designed in various shapes, some with glowing flowers and leaves—when he was younger, he had even blown his own glassware. The whole place stank of ozone (I came to love that smell of happy toxicity) as he shot crackling blue sparks from gun-shaped electrodes to illuminate the little neon lights. I found this world completely entrancing—the beautiful coils of thin copper wire I could make out in the innards of his electrical meters, the beautifully finished brass contact posts mounted on black Bakelite stands.

Once a week, on Saturday, he would go to the venerable patent room at the New York Public Library, where he would sift through American, French, and German publications to see what had been done in his area. Then it was back to the laboratory to invent something brand-new.

All this struck me, at the time, as of a piece with the imagined technological future, the science that everyone was talking about but that wasn't quite accessible to me: rocket countdowns on television, IBM computers in the glossy magazines. But my great-grandfather's basement laboratory was actually very much in the past. I'm certain that he and transistors never met. His desk upstairs, too, with its neat stack of precision technical drawings bearing his carefully written marginal commentary in fountain-pen ink, was more reminiscent of pre–World War I Europe. Or maybe, to be more precise, what I was seeing was the future of the past—of the electric lights and trans-formers that in Edison's New Jersey laboratory had prom-ised something unclear but new and hopeful for the world after 1900. Whatever it was, I was crazy about it—and by extension, about everything electrical.

One day when I was eleven or so, I made a tic-tac-toe computer, the whole thing wired somewhat chaotically with switches and bulbs onto some scrap lumber. By then, my great-grandfather was just about blind. He was sitting in his garden chair by the bird feeder, and I told him what I had done with the copper strips and the screws and the yards of orange-colored wire. He asked me to show it to him. Slowly, my own Albert Einstein ran his fingers along the wires, testing to make sure the connections were good, and explained to me very precisely how I could simplify it.

In seventh grade, I had a science teacher who, unlike any of my previous teachers, actually knew some physics. I told him about my fascination with electricity and magnetism, and he showed me, step by step, how Einstein, using basic ideas about the speed of light and coordinated clocks, had concluded that a moving bar would be measured as shorter when compared with a bar at rest and how moving clocks went slower than motionless ones. It was a stunning moment for me. I had managed to get a ham radio license, but I had never seen anything interesting derived in physics. This was magic! I copied over the argument many times, slowly, like the words of a prayer. I can still picture my Δx's and Δt's in heavy pencil on the lined pages of my notebook.

In retrospect, it seems so woefully partial. I didn't know any classical physics—in fact I knew just about nothing of science other than how to take apart (and sometimes even fix) the broken vacuum-tube radios I cadged from local repair shops. But I thought that this argument of Einstein's was the most astonishing thing I had ever heard: you could begin with simple assumptions and figure out something completely unexpected about the world.

I fell in love with physics. (Much to my later regret—indeed, it should never have been allowed—I managed to learn almost no biology.) It was politically difficult to hold on to a dream of physics in high school, with the country in the middle of the Vietnam War. My friends considered hard science a short step from the manufacture of plastic shrapnel. I took physics by myself, after hours, from a sympathetic teacher—samizdat pulleys and covert capacitors. I ended up graduating early, and spent the following year in Paris, where the École Polytechnique let me work with a wonderful researcher in a plasma physics laboratory and audit a math course in distributions and convolutions given by the great mathematician Laurent Schwartz. Again—and indeed still today—I was riveted by the fact that the symbols on the page, those abstract musings, somehow linked up with the oscilloscopes, copper wires, and machines on the laboratory floor.

This contact point between abstraction and concreteness has remained a central theme of my work. As a student in Paris, I started reading Einstein's papers, pursuing that union of machine reasoning and abstract concepts I found so compelling about his work. It was not the celebrated Einstein who interested me; it was quite distinctly the younger Einstein, the Einstein who grew up around his father's and uncle's electrotechnical company, the Einstein who spent his university days messing with experiments in the basement, ducking out of the remarkable mathematical lectures given by greats like Hermann Minkowski. When I returned to the United States and started college, I read Thomas Kuhn's *The Structure of Scientific Revolutions* and Gerald Holton's *Thematic Origins of Scientific Thought;* both showed me a new side of Einstein's work, one connected to history and philosophy. Those books extended the connections and opened up for me the possibility of thinking about Einsteinian physics in a completely new way.

What now seems to me something of a minor obsession continued during my undergraduate years; for one summer, at the Institute for Advanced Study in Princeton, I worked on the Einstein Papers publication project, which was just then getting under way. I found it extraordinary to see how deeply Einstein had been engaged with detailed discussions of inventions and patents. For my Ph.D. thesis in the history of science (which became my first book, *How Experiments End*), I used the case of Einstein's work on the gyrocompass—a nonmagnetic way of tracking one's orientation—to show how technological concerns, the grit of the basement, lay behind some of Einstein's most abstract thought experiments. The gyrocompass became for Einstein a model of the atom. Pure physics met applied engineering.

This preoccupation led many other places. I became fascinated by detectors, those machines that translate the invisibly small into a larger world where their interactions match up against the great synthetic accounts of high-energy theory. My great-grandfather's laboratory bench—and, later, Einstein's—spurred an interest in laboratory architecture. The electrical universe I had glimpsed long ago in that New York City laboratory and the patent papers of Einstein have led, more recently, to an examination of the ways in which Einstein and Henri Poincaré, in different ways, used the idea of clock coordination as they formulated their ideas on the relativity of time.

Much as I admire the older Einstein—much as I find him an admirable figure for his political courage in opposing McCarthyism, nuclear escalation, and racism, much as I see the bravery behind his pursuit of a unified field theory—it is the younger Einstein who has meant the most to me. There is a turn in Einstein's later life and work that (though I sympathize with it) breaks its connection with me—a move away from the engagement with things and thoughts that characterized him as a younger scientist. I don't think this change of focus was purely intellectual. I think the Nazis' rise to power and his exile from Europe were deeply traumatic to Einstein—more so than is evident from his public pronouncements. After the Holocaust, he found it impossible to reestablish contact with Germany, but not only with Germany. I think that in a certain sense he stepped back from the world. It is as if from the intimate connection of things with thought, only thought survived. Of course, the horror in Europe was not the only reason for his farewell to the laboratory; idiotically, the American security services considered him a threat and excluded him from the important war work (though he did do some work on the theory of torpedoes). No doubt, too, his increasing fame created an awkwardness in his interaction with other physicists. The Institute for Advanced Study gave him the peace he craved, a place where he could talk at length with a few select friends, especially Kurt Gödel. Yet there was a very great distance between those long walks in exile and the turbulent, productive chaos of his early years, those years of the basement lab, when he was experimenting on magnetism, working on general relativity, exploring quantum mechanics, and testifying about gyrocompasses.

When I think about Einstein, I always come back to my great-grandfather's laboratory, to those thin tubes of neon we lit with the crackle of a spark. The vestiges of 1900 were in his fingers as he showed me how to design circuits and make things work. Similarly, in all of physics there is nothing as beautiful to me as the simple, principle-based reasoning of the young Einstein while he sorted out the relativities and the quanta, always thinking about knobs and tubes. In Einstein's early prose, there is a clear and luminous reasoning that is never far from the world we can touch. He imagines a man falling from a roof, his tools falling with him—and in that moment he grasps the principle of equivalence. After all these years, I still find the physicality of thought, the abstraction of the material world, utterly captivating.

Childe Bernstein
to Relativity Came

JEREMY BERNSTEIN

JEREMY BERNSTEIN is emeritus professor of physics at the Stevens Institute of Technology. He was also for thirty years a staff writer for *The New Yorker,* for which he wrote profiles of leading physicists, including, besides Einstein, I. I. Rabi, Hans Bethe, and J. S. Bell. He has also been an adjunct professor at Rockefeller University and is an honorary trustee of the Aspen Center for Physics. He has held appointments at the Institute for Advanced Study, the Brookhaven National Laboratory, the European Organization for Nuclear Research (CERN) in Geneva, Oxford University, the University of Islamabad, and the École Polytechnique in Paris. He is the author of some twenty books, including *Albert Einstein and the Frontiers of Physics, Hitler's Uranium Club,* and most recently *Oppenheimer: Portrait of an Enigma.*

In the late 1930s, when I was growing up in Rochester, New York, small numbers of Jewish refugees were coming to the city. They were the lucky ones, able to get out of Europe and into this country. These were people with considerable professional skills, who could find work. My father, a prominent Reform rabbi in Rochester, helped them to settle in. Among those he helped were two physicists, Victor Weiskopf and Max Herzberger. Victor (Vikki to the physics world) was a protégé of Niels Bohr's and would later become the star of MIT's Physics Department. Vikki has claimed that he met me in 1937 when he came to Rochester, but since I was then seven years old it did not make much of an impression. Max Herzberger was something else. He was one of the world's experts in geometrical optics and had written the volume on the subject in the famous Springer "Yellow Peril" series (so called because the volumes were bound in yellow jackets and notoriously difficult). He had also been a lens designer for Zeiss; he now became a lens designer for Eastman Kodak, which was, of course, located in Rochester. Max had taken his degree in Berlin, and Albert Einstein had been on his Ph.D. examining committee. Max maintained a relationship with Einstein that endured even though they were on separate continents.

Max loved explaining the theory of relativity to anyone, under any circumstances—in fact he was usually oblivious of the circumstances. Sometimes he would switch into German, apparently unaware that he had done so and that his listener did not speak the language. On one famous occasion, he was observed relentlessly explaining the theory to an elegantly dressed Rochester matron at a garden party, heedless of the fact that her little terrier was peeing on his shoe. When I was in high school, Max tried to explain relativity theory to me. I had no clue as to what he was talking about and no interest. Years later, when I was seriously studying physics, he explained to me some of his ideas about quantum mechanics. They were odd. He objected to the use of complex numbers, arguing that anything you measure has a real number value, so the theory should contain nothing but real numbers. He tried this out on Einstein. He showed me Einstein's brief response. Translated, it said, "Since I do not understand your problem, I also do not understand your solution."

Einstein could be dismayingly blunt—just read some of his letters assessing people who were looking for physics jobs. In the late thirties, my father had an illustrative encounter. With his teacher Stephen Wise, he attended a meeting at which Einstein gave a talk. At some point, my father said to Wise, loudly enough for Einstein to hear, "I am giving a sermon about how I met Einstein and I have never met Einstein!" Einstein spoke up: "Rabbi, you have lied to them before, now you will lie to them again."

In high school, physics was one of my worst subjects. I was pretty good in math, but if anyone had told me that there were actual living mathematicians—as opposed to, say, Euclid and others who had created the stuff—I would have thought they were crazy. You might well ask how, with this background, I ever became a physicist myself, let alone the author of a number of articles on Einstein and an entire book in which I attempt to explain both the man and his theories. (Indeed, it seems to me that I have spent much of my professional life trying to explain Einstein and his theories.) In some large sense, I owe this transformation to James Bryant Conant, who in my undergraduate days at Harvard was president of the university. Conant had spent World War II helping to direct the mobilization of science for the war effort. He came out of the war persuaded, especially by the development of nuclear weapons, that science was too important to be left to the scientists. Thus he brought about the General Education program at Harvard, with its requirement that every undergraduate take a science course in order to graduate. (There was also a requirement that you be able to swim two laps in the pool; the former requirement was to enable Harvard graduates to think sensibly about science-related matters, while the latter, I suppose, was to keep you from drowning if you happened to fall off a yacht.) Thus it was that when I arrived in Cambridge in the fall of 1947 I was confronted with choosing which of the General Education science courses I should take.

There was an aid: the *Confidential Guide* to the courses, a candid and irreverent survey by upperclassmen that enabled me to discern which of the Gen Ed science courses was the easiest. Clearly, this was Natural Sciences 3, taught by the late I. Bernard Cohen, a well-known historian of science. Very much later, when I was a graduate student, I became a teaching assistant for Cohen's course. By then I had come to realize that where modern physics was concerned, Cohen was something of a fraud. He understood rather little, which did not keep him from being an excellent teacher for undergraduates who, like myself, understood essentially nothing. Most of Natural Sciences 3 dealt with subjects Cohen really did know quite a lot about—science, that is, from the Greeks to the end of the nineteenth century. Cohen was a Newton scholar, so Newton's ruminations were a highlight of the course.

Then we came to twentieth-century physics. Up until that point I had been sleepwalking, but I suddenly awoke with a start. I no longer remember which topic came up first, relativity or quantum theory, but since the course was historical I imagine it was relativity. What caught my attention were the theory's apparently absurd predictions. I remember my bemusement when Cohen explained that to a stationary observer a uniformly moving object appears to gain mass, and that this gain in mass becomes infinite as the object approaches the speed of light. I realized that it did not matter whether the object was at rest and the observer was in motion or vice versa: there was the same mass gain. So . . . you could impart mass to an object just by running past it! This idea was especially troubling because Cohen had given us a misleading definition of mass in his treatment of Newtonian physics—that is, mass as the quantity of matter. Since matter was atomic, it seemed—falsely—that relativity predicted that the number of atoms increased when an object was in motion. What also caught my attention was Cohen's claim that only ten, or fifteen, or some other small number of people understood the theory. I imagined he included himself, and surely Max was one of the elect.*

---

*It was only later that I understood that "mass" means something else in relativity and what the origin of the notion that relativity was understood by almost no one was. In 1923 the British astronomer Arthur Eddington published a classic monograph on relativity. When he was asked if it was true that only three people understood the theory, he responded, "Who is the third?"

After Cohen made this claim, I decided, more or less for the heck of it, that I would become the eleventh, or the sixteenth, or whatever, person to understand the theory. It did not occur to me to look at the course catalog of the Physics Department, where, needless to say, I would have found that relativity was routinely taught. I simply took Cohen's remark at face value. I also did not have any idea what it meant to "understand" a physics theory like relativity. The kind of understanding I was familiar with from high school involved being able to translate a foreign language like Latin into English; having done that, one understood the Latin. Understanding geometry meant being able to repeat the steps of a proof on an exam. Understanding a poem meant understanding, perhaps with the aid of a dictionary, all the words and allusions in it. I once brought one of my assigned poems home; my father read it and asked if I agreed with what it said. The question completely threw me, as it had never occurred to me that poems said anything you could agree or disagree with— they were just words. I assumed that understanding relativity was something like this. I would find a book and, with the aid of a dictionary, translate all the unfamiliar words into ones that I understood. I was prepared, if necessary, to devote a couple of months to this project. So I went off to Widener Library to look for a book, preferably one by Einstein, since he obviously understood the theory.

There are those who have said that of all the Einstein books in Widener, I made the worst choice. But in retrospect I have to disagree. The book I chose—because of its title—was his *The Meaning of Relativity*. This was the text of the Stafford Little Lectures that Einstein gave in Princeton in 1921. Since it was 1948 when I took the book out of the library, it must have been the second edition, published in 1945. I have in front of me the third and last edition, published in 1950. The 1945 edition had a remarkable appendix in which Einstein disowns what he calls the cosmological constant, which he had introduced in 1917 in order to keep the universe—then the Milky Way galaxy—stationary; within a decade, Edwin Hubble had shown that the universe was expanding, and Einstein changed his mind. For contemporary cosmologists, the cosmological constant has now come back with a vengeance. In any event, the text itself was about a hundred pages long. I thought that if I could read two or three pages a day, I would get the job done in the time I had allotted.

The first three pages went fairly nicely. He begins by saying, "The theory of relativity is intimately connected with the theory of space and time." This sentence had a nice ring to it and was followed by some philosophical remarks. I was not sure why they were important, but I thought I understood what they meant. Thus ended the first day. If it continued like this, it would be a piece of cake. But on the second day, I ran into the equation

$$\Delta x_v' = \sum_\alpha \frac{\delta x_v'}{\delta x_\alpha} \Delta x_\alpha + \frac{1}{2} \sum_{\alpha\beta} \frac{\delta^2 x_v'}{\delta x_\alpha \delta x_\beta} \Delta x_\alpha \Delta x_\beta \ldots$$

Thus ended my attempt. I had hit a wall, and no dictionary was going to see me through. I went to Cohen for help. That was the important step. If I had taken out one of

Einstein's more popular books, I might have muddled through, with the illusion that I (sort of) understood it. But having chosen *The Meaning of Relativity*, I saw that I was out of my depth and I needed help. Cohen made a suggestion that changed my life.

He said that in the spring semester there was another Gen Ed course at about the same level as his but concentrated on modern physics. It was taught, he said, by Philipp Frank, who had just published a biography of Einstein and moreover had succeeded Einstein as the theoretical physics professor at the German University of Prague in 1912, when Einstein moved to the Eidgenössische Technische Hochschule in Zurich. This course sounded like just what I wanted, so I signed up at once.

On the first day, Frank appeared in the large lecture room in the physics building. He was a short, almost bald man, with the face of a very intelligent basset hound. He walked with a limp that dated to an encounter with a streetcar in Vienna, where he was born in 1884. His accent was difficult to place; I later decided that the ten or so languages he spoke were piled on top of one another like the cities of Troy, with various remnants poking through here and there. When I eventually got to know him, he told me that in the 1930s in Prague there had been three factions— the Nazis, the Communists, and the Jews. The Nazis were afraid that the Russians would invade, the Communists were afraid that the Germans would invade, and the Jews were afraid of everybody. The only thing they could agree on, he said, was to hire an English teacher so that they could all emigrate to the United States.

His lectures were remarkable. He could take the most complex subject and reduce it to a language so simple that you were sure you understood it. It was only later, when you tried to reconstruct the arguments, that you saw how subtle they were. A few years ago, two colleagues and I published a textbook in modern physics. I wrote the relativity section, and to do so I unearthed the notes I had taken in Professor Frank's course in the spring of 1948. Every so often in the course, Professor Frank would make a more technical aside, which he would begin by saying, "If you know a little of mathematics . . . " It became clear to me that if I was really going to understand this, I had better learn "a little of mathematics." I ended up majoring in it. By the time I graduated, I had gotten to know Professor Frank and his wife, Hania, rather well. I learned that he had been a student of the great Austrian physicist Ludwig Boltzmann, whom he once told me was the most gifted mathematical physicist he had ever known. I learned that when Frank was a young man he was as interested in the philosophy of science as he was in physics and later became a founding member of the Vienna Circle, the creators of what became known as Logical Positivism. I once took a reading course with him in which we read Wittgenstein's *Tractatus* together. I learned that while he was in Prague he was part of an intellectual circle that included Franz Kafka and Kafka's biographer, Max Brod. When Professor Frank retired, I helped him clean out his desk in the Physics Department. There were letters from such renowned physicists as Erwin Schrödinger that he had not bothered to open. We opened one from Schrödinger, which began

"Just between us daughters of parsons . . ." and went on with some gossip that had been fresh several years earlier. "You see, it wasn't so important anyway," Professor Frank remarked. He died on July 21, 1966, and I was one of the speakers at his memorial service.

By the end of my sophomore year, I felt I understood enough of at least the philosophical background of relativity that a chat with Einstein might be helpful. Professor Frank raised this lunatic idea with Einstein. In the spring of 1949, I received the following, sent from 112 Mercer Street in Princeton. It was dated June 3 and reads:

Dear Mr. Bernstein,

I am sending you enclosed paper in which I expressed opinions from an epistological [*sic*] point of view. I do not give oral interviews to avoid misinterpretation.

It was signed "Sincerely yours, A. Einstein."

The Books in the Basement

GEORGE JOHNSON

GEORGE JOHNSON writes about science for the *New York Times* from Santa Fe, New Mexico. In 1999 he received the AAAS Science Journalism Award. He is also a codirector of the Santa Fe Science Writing Workshop. Johnson's books include *Fire in the Mind: Science, Faith, and the Search for Order; Strange Beauty: Murray Gell-Mann and the Revolution in Twentieth-Century Physics; A Shortcut Through Time: The Path to the Quantum Computer;* and most recently *Miss Leavitt's Stars.*

Early in my college career, I was perusing the science section of my favorite bookstore in Albuquerque—the Living Batch, where the really smart hippies hung out—when my eye was caught by the spine of a little paperback called *The Universe and Dr. Einstein*. Priced at ninety-five cents, it promised to be "the clearest, most readable book on Einstein's theories ever published." On the cover was a tantalizing portrait of the great scientist, his wild shock of hair blowing in the cosmic wind. Behind him loomed the night sky, shining with constellations and mathematics. This was clearly the man who knew the answers, and they would be imparted to me, a mere humanities major, in a book only 118 pages long. I bought it on the spot.

It seemed extraordinary that such a document existed. Written by a journalist for *Life* magazine named Lincoln Barnett, it had appeared (according to the fine print on the copyright page) in shorter form in *Harper's*. Flipping quickly through the pages, I saw to my relief that it was filled with prose, not equations. There was some scary-looking algebra way back in an appendix, but I figured that by the time I had breezed through the little book, even the math would be clear. Most impressive of all, this compact artifact of scientific exposition was recommended, in a one-page foreword, by Einstein himself. I probably didn't appreciate it at the time, but what he had written was a miniature essay, just three paragraphs long, on what makes good science writing.

"Anyone who has ever tried to present a rather abstract scientific subject in a popular manner knows the great difficulties of such an attempt," Einstein wrote.

Either he succeeds in being intelligible by concealing the core of the problem and by offering to the reader only superficial aspects or vague allusions, thus deceiving the reader by arousing in him the deceptive illusion of comprehension; or else he gives an expert account of the problem, but in such a fashion that the untrained reader is unable to follow the exposition and becomes discouraged from reading any further.

If these two categories are omitted from today's popular scientific literature, surprisingly little remains. But the little that is left is very valuable indeed.

He was recommending Mr. Barnett's book as one that had steered a steady course between the shoals. It was crucial, Einstein observed, that works like this be published: "Restricting the body of knowledge to a small group deadens the philosophical spirit of a people and leads to spiritual poverty."

Recently, I retrieved my old copy from a box in the basement of my childhood home and started rereading, experiencing all over again the excitement of confronting Einstein's science for the first time. Lincoln Barnett may not have been my first cut through the brambles of relativity and quantum mechanics. I've also unearthed a yellowed copy of Bertrand Russell's *The ABC of Relativity*, and I may have dipped into George Gamow's *Mr. Tompkins in Wonderland*. But I'm pretty sure it was *The Universe and Dr. Einstein* that first made the concepts come alive—and made me realize that a person could approach this world of ideas not just as a scientist but as a writer.

From the first sentence of Barnett's book, you know you are in good hands: "Carved in the white walls of the Riverside Church in New York, the figures of six hundred great men of the ages—saints, philosophers, kings—stand in limestone immortality, surveying space and time with blank imperishable eyes." Among them, of course is Albert Einstein, "the only one who shook the world within the memory of most living men." Alas, Barnett lamented, hardly anyone outside the world of physics had more than the dimmest notion of just what Einstein had done. Here it was nearly half a century after his first astonishing papers (Barnett's book was first published in 1948) and the ignorance stubbornly persisted: "Today most newspaper readers know vaguely that Einstein had something to do with the atomic bomb; beyond that his name is simply a synonym for the abstruse." And it was not just the uneducated who were missing out. As Barnett put it, "Many a college graduate still thinks of Einstein as a kind of mathematical surrealist rather than as the discoverer of certain cosmic laws of immense importance in man's slow struggle to understand physical reality."

I'd been oscillating since freshman year between two poles, majoring in literature one semester and physics the next. I tried to pay attention as our stately professor, Dr. Victor Regener, led the way through Newton's laws, rolling out the inclined planes and frictionless tracks to drive home the point that things really moved as the equations described. I struggled through the early chapters of the thick blue brick we called "Halliday & Resnick," an albatross of a textbook that I would tote in the evenings to the Casa Luna Pizzeria, where I drank coffee, flirted with the waitress, and tried to solve the problems at the end of the chapter:

A dog is looking out a second-story window when a ball bounces up from the street, passes the top of the window frame and returns one second later on its way back to the ground. If the window is 15 feet above the pavement, then how old is the dog?

Or something like that. It was time for a third cup. Going through the physics course descriptions in the university catalog, I realized that by my senior year I would be all the way up to the nineteenth century. (I think a survey of relativity and quantum mechanics was offered as an elective.) Only many years later, when I'd earned a Ph.D., would I be taken into a chamber where, like a thirty-third-degree Freemason, I'd see the true mysteries revealed—the shrinking rulers and the slowing clocks . . . and why all this made E equal mc^2.

Or I could sign up for "Literature of the Beat Generation" and read Barnett after class. For me, that was the right decision. Chapter 3, page 23, and I was already learning a little about Max Planck and the quantum, a prelude to Einstein's photoelectric effect. That led to a short tangent on wave–particle duality, with a little Schrödinger, Heisenberg, Bohr, and Born thrown in. Fifteen pages later and Barnett was laying the foundation for special relativity: the traveler strolling on the deck of the moving ship, the surprise of the Michelson-Morley experiment, the two trains and the lightning bolt . . . and there in a footnote were the curious zigzags of the Lorentz transformation. The math wasn't so scary after all. You could actually see, with just a little algebra, how as something approaches the speed of light, time stands still, length goes to zero, and mass becomes infinite. No wonder you could go no faster, that there could never be the optical equivalent of a sonic boom.

By Chapter 9, I was immersed in the "four-dimensional space-time continuum," riding Einstein's plunging elevator and watching the bending flashlight beam—encountering the rest of the pedagogical furniture of relativity that is hauled on stage by science writers again and again. Matter bends space, and space tells matter how to move. I was amazed that I could, sort of, understand this stuff. Maybe what I was experiencing was closer to what Einstein called "the illusion of comprehension," but that was OK—all I was looking for was a toehold, something that would let me climb a little higher, reach for another rung.

Later on, I would encounter these ideas again, in another first-rate popularization, Barbara Lovett Cline's *Men Who Made a New Physics*. (I just opened up my old copy and found a letter from the pizzeria waitress, postmarked from Mexico, marking the chapter on Einstein's miraculous year.) After college, as I covered the police beat for the *Albuquerque Journal,* I tried to burrow deeper into the literature of relativity with some of the scientists' own more or less popular accounts: Arthur Eddington's *The Nature of the Physical World,* Einstein and Leopold Infeld's *The Evolution of Physics,* Edwin F. Taylor and John Archibald Wheeler's *Spacetime Physics.* I imagined myself at the apex of a light cone and pondered the notion that it's not just the speed of light that's absolute but the speed of signaling—that a rational world of cause and effect requires that you cannot learn of an event before it happens. Saying that Einstein proved that "everything is relative" is exactly wrong, for he identified the standard that makes comprehension possible. As he and Infeld put it, if human beings could break the electromagnetic speed limit, "We could see occurrences from the past by reaching previously sent light waves . . . catch them in a reverse order to that in which they were sent, and the train of happenings on our earth would appear like a film shown backward, beginning with a happy ending." A truly weird universe would be one without relativity.

These insights—the insights of an amateur—fade from disuse, only to be rekindled every few years as I open a new book on Einstein and take in another production of the metaphorical stage play. The trains and the lightning bolts, the elevator and the light beam—coming upon them is like encountering old friends. With each retelling, the ideas settle in a little more comfortably.

Sometimes there is even a fresh metaphor to appreciate. The parable in João Magueijo's *Faster Than the Speed of Light* about Einstein, cows, and an electric fence made the illusory nature of simultaneity clearer to me than ever before. And a couple of pages in the second chapter of Brian Greene's *The Elegant Universe* induced nothing less than an epiphany: c is not just the speed of light and the speed of signaling but the speed at which everything in the universe is moving through the space-time continuum. Wow!

Greene asks us to imagine a racing car traveling at a fixed speed across a flat expanse. Its velocity is divided between two components, north–south and east–west. The faster it moves in one direction, the slower it must move in the other—a zero-sum game. An airplane divides its speed among three dimensions, and Einstein is just asking us to add one more: in the relativistic universe, all motion is shared among four dimensions. As I sit at my desk going nowhere, I am moving full speed ahead through time. If I get up and start walking, my spatial velocity must be subtracted from my temporal velocity. My watch runs incrementally slower and I don't age quite so rapidly.

The faster you move through space, the slower you move through time. Confronting this idea from a new (for me) perspective jogged loose a memory of my favorite Robert A. Heinlein novel from junior high school, *Time for the Stars,* which revolves around the famous twin paradox. The brother who boards a starship as a boy returns home a few years later to find that his double is now a very old man. Was it really true, I had wondered back then, that a very smart person named Albert Einstein had proved scientifically that such an absurdity was possible?

But now the idea doesn't seem so crazy. I can almost feel myself existing, like a star or an electron, as a ripple in four-dimensional space-time. For a writer looking for material, it doesn't get any better than this.

How He Thought

LEONARD SUSSKIND

LEONARD SUSSKIND is the Felix Bloch Professor of theoretical physics at Stanford University. His contributions to physics include—besides the discovery of string theory, black-hole complementarity, and the holographic principle—work in quark confinement, the internal structure of hadrons, Hamiltonian lattice gauge theory, symmetry breaking, baryon production in the universe, quantum cosmology, and M-theory. He is the coauthor (with James Lindesay) of *An Introduction to Black Holes, Information and the String Theory Revolution: The Holographic Universe* and the author of *The Cosmic Landscape*.

Einstein rescued me from having to spend the rest of my life as a plumber in the South Bronx. My father, Ben Susskind, was a plumber, and his dream was that he and I would someday be in business together, fixing lead pipes and cleaning sewers. Benny was the heroic figure in my life—a tough man who of necessity had had to leave school in 1917 at the age of twelve to go out and make his living in the dirt and grime of rat-infested tenement buildings. I was well acquainted with that life. From the age of fifteen, I had worked as a plumber's helper and then as a journeyman plumber during the summers and weekends, and sometimes late at night. I hated it, but until I was twenty-one I had no other ambition.

I had always loved mathematics, but it wasn't until I started CCNY, the poor people's college in Harlem, that I found out there was something called physics. I fell in love with it. The only problem was that I was going to have to tell the old man I didn't want to be a plumber. So one evening after work, I screwed up my courage and took my wife and baby over to my parents' house. I found him in his shop cutting some pipe for the next day's work, and with great fear in my heart I said, "Ben, I'm not going to be a plumber."

He looked me up and down and said, "What the hell do you mean, you ain't gonna be a plumber?"

"I'm not going to be a plumber," I repeated.

"Then what the hell are you gonna be? A ballet dancer?"

There and then I told him I wanted to spend my life as a physicist, a word that meant nothing to him. "Physicist? What's a physicist?"

I told him it was a kind of scientist. Thinking he might have thought I wanted to be a pharmacist, I explained that Albert Einstein had been a physicist. Ben's countenance turned thoughtful: "Einstein? Are you any good at this stuff?"

I told him I thought I could be very good at it. He pondered this for a moment or two and then gently poked me in the chest with a short piece of pipe. "No plumber! You can be a physicist if that's what you want."

My mother, who had heard the news from my wife, came into the shop distraught. She was crying. We would have no money, she said, and the baby would starve! Ben turned around and shot her a piercing glance. "Shut up! He's gonna be Einstein!"

For most, the Einstein of their imagination is the wise old man with the unruly hair and the sad, drooping eyes, the Einstein who had seen it all: World War I, the Nazi takeover, the start of the Manhattan Project—the iconic Einstein. That's not my Einstein. I'll take the young, dapper, almost elegant twenty-six-year-old—the Einstein of 1905.

The stylish suit and the well-trimmed mustache have little to do with it: it's the mind that fascinates me—a way of thinking that I admire above all others. In that year, whose hundredth anniversary we have just celebrated, Einstein was at the height of his powers. He had an almost supernatural way of looking into nature and seeing clearly what others could see only as cloud-shrouded shadows. Not that he could decipher unusually complicated formulas, digest difficult mathematics, or remember prodigious amounts of experimental information. Einstein's style was to begin with the simplest observations about nature—things so simple even a clever child could understand them. But from these elementary considerations, he drew the most profoundly far-reaching conclusions. The things he saw were in retrospect obvious, but no one else had seen them.

Take, for example, the first paper on relativity theory. Other physicists, like the mighty Hendrik Lorentz, knew all the things the twenty-six-year-old patent clerk knew. But they could not free themselves from old baggage: the view that light and other electromagnetic phenomena were disturbances of the aether, a hypothetical transparent substance that filled space and whose stresses, strains, and vibrations were the electromagnetic fields of Faraday and Maxwell. (I always think of the aether as clear Jell-O.) Other physicists knew that no experiment had succeeded in detecting the earth's motion through the aether, so they concocted complicated explanations of how the aether deformed objects such as meter sticks just so it would be undetectable. By contrast, Einstein's paper "On the Electrodynamics of Moving Bodies" (*Annalen der Physik,* September 26, 1905) begins with an observation so simple that anyone can understand it. Translated from German, the first paragraph reads:

It is known that Maxwell's electrodynamics—as usually understood at the present time—when applied to moving bodies, leads to asymmetries which do not appear to be inherent in the phenomena. Take, for example, the reciprocal electrodynamic action of a magnet and conductor. The observable phenomenon here depends only on the relative motion of the conductor and the magnet, whereas the customary view draws a sharp distinction between the two cases in which either the one or the other of these bodies is in motion. For if the magnet is in motion and the conductor at rest, there arises in the neighborhood of the magnet an electric field with a certain definite energy, producing a current at the places where parts of the conductor are situated. But if the magnet is stationary and the conductor in motion, no electric field arises in the neighborhood of the magnet. In the conductor, however, we find an electromotive force, to which in itself there is no corresponding energy, but which gives rise—assuming equality of motion in the two cases discussed—to electric currents of the same path and intensity as those produced by the electric forces in the former case.

Stripped of its German linguistic contortions, Einstein's point was simple: Take in your left hand a coil of wire and in your right hand a magnet. According to nineteenth-century physics, if you move the coil through the aether, holding the magnet fixed, the magnetic field of the magnet will push on the electrons in the moving coil and cause a current to flow. But if, instead, you hold the coil fixed and move the magnet through the ether, the moving field of the magnet creates an electric field that pushes the electrons through the stationary coil. Two completely different phenomena (or so it seemed), one involving magnetic fields and the other electric fields, yet the results are exactly the same: electric currents flow through the coil. What was Einstein's viewpoint? The cause of the current is the *relative* motion of conductor and magnet, not the motion of the magnet through the aether or the motion of the coil through the aether. From this and his equally simple study of the synchronization of clocks by light signals (so wonderfully explained in Peter Galison's 2003 book, *Einstein's Clocks, Poincaré's Maps*), Einstein deduced the special theory of relativity and the fateful formula $E = mc^2$.

Soon after the annus mirabilis of 1905, the young man with the sad eyes began to think about gravity. The obvious route was to try to modify the equations of electrodynamics in some way so that Newton's law of gravity would emerge as a first approximation. Eventually this might have succeeded, but only after vastly complicating the equations. I have little doubt that most physicists would have proceeded in this way, but not Einstein. Instead he thought about elevators. Einstein's point of departure was an elementary observation that anyone who has ever been on a high-speed elevator will recognize: as the elevator starts its upward journey, the acceleration makes it feel as if gravity had temporarily become stronger; moreover, as the elevator comes to a stop, gravity seems to be suspended for a brief time. Einstein postulated that the equivalence of gravity and acceleration is a fundamental law of nature and called it the equivalence principle. Others had known this connection—even Newton. But only Einstein had the clarity of vision to follow this idea for the next decade until he finally produced the greatest theory of all—the general theory of relativity. The general theory is not simple: the mathematics is difficult and the equations are complicated. But it was from the humble elevator thought experiment that Einstein derived the basic idea. For me, this way of thinking represents the highest form of beauty that science can offer.

One of the most striking things about Einstein's thinking was how sure-footed he was in those early days. He knew exactly what to trust and what to discard. How could he have been so certain that the laws of thermodynamics and their statistical interpretation were more trustworthy than those of classical mechanics and the wave theory of light? But without that confidence, he probably would not have discovered that light is composed of indivisible quanta of energy—the particle-like photons. What made him certain that the velocity of light was universal but that the concept of universal simultaneity should be abandoned? How did he know to trust his senses about elevator rides but not about the addition of velocities? I suppose we can say only that Einstein's mind and instincts were particularly well tuned to the subtleties of the physical world.

At this writing, I am sixty-four years old and somewhat inclined to look back on my own scientific accomplishments. In some instances, I am a little embarrassed that I didn't see as far as I should have. Some accomplishments, like the discovery of string theory, are deeply satisfying. But the best of all are the few times when I feel that I was thinking, if only in a small way, in the style of Einstein. Perhaps the strongest feeling of that kind has to do with the discovery of a principle—the principle of complementarity in black-hole physics.

In 1976 Stephen Hawking raised a profound question about gravitation and its relation to quantum mechanics. Hawking's observation was that quantum mechanics causes black holes to eventually evaporate, leaving only thermal radiation (photons) in their place. But what of the objects that fell into the black hole—is all the information about them obliterated by falling behind the black-hole horizon?

This was a brilliant question. The standard laws of quantum mechanics forbid information's being irrevocably lost—scrambled, yes; lost, no. On the other hand, the laws of general relativity say that the horizon is a point of no return; any information—such as that in a telephone book, a computer, or a human genome—that falls through the horizon cannot get back out; therefore, apparently, it will be permanently lost. In other words, we have a paradox: the laws of gravity and the laws of quantum mechanics appear to be incompatible.

Here is a graphic example. Suppose a friend of yours falls into a giant black hole a billion light-years in diameter. According to the general theory of relativity (really just the equivalence principle), nothing special would happen to your friend as she fell past the horizon. Her body would eventually be destroyed perhaps a half billion years later at the very center of the black hole—the infinitely violent singularity—but the horizon crossing would be a non-event in her experience.

But quantum mechanics and statistical mechanics lead to a very different story. You, who are outside the black hole, reconstruct the story by examining the eventual evaporation products of the black hole, and what you find is gruesome. Your friend became hotter and hotter, not at the singularity but at the horizon. She became so hot, a million trillion trillion degrees, that she was turned into thermal radiation and eventually came back out as photons. Surely these histories cannot both be correct!

Hawking trusted the wrong thing. He trusted ideas inherited from classical general relativity. It seemed plain to him, and to most people, that information that went behind the horizon would be lost forever. But that meant that quantum mechanics, as well as the statistical interpretation of thermodynamics, must be wrong for black holes.

To me, and to the Dutch physicist Gerard 't Hooft, this seemed incorrect. Giving up the basic principles of statistical mechanics and thermodynamics was too high a price to pay. I began to devise thought experiments, of the kind that Einstein used so well. Imagine that you have dropped an electron into a black hole . . . and so on. After analyzing many of these imaginary experiments, I concluded that there was no paradox. Both histories are true! But there is a completely new relativity principle that goes far beyond Einstein's—the principle of black-hole complementarity. Without getting into technical details, I can describe the principle as holding that in a certain sense information can be in two places at once. More accurately, the location of an event or a bit of information is not invariant; different observers may not agree about where or when your friend met her grisly fate. Much as Einstein concluded that the simultaneity of events is relative, I concluded that whether an event takes place behind or in front of the horizon depends on the state of motion of the observer.

In time, another principle began to emerge from thought experiments. The so-called holographic principle put forward by 't Hooft and myself may be the strangest twist in physics since the early days of quantum theory and relativity. Given that things may not be where they seem to be, 't Hooft and I were led to conclude that our universe is a kind of hologram, a two-dimensional representation of everything in the three-dimensional world. Once thought to be a wild conjecture, the holographic principle is now a pillar of modern theoretical physics. Rightly or wrongly, I like to think that the young Einstein would have approved.

Einstein once said, "The essential thing in a man like me is what he thinks and how he thinks, not what he does or suffers." He might also have said, "The greatest satisfaction is not in what I found but in how I found it."

Toward a Moving Train

JANNA LEVIN

> JANNA LEVIN is a theoretical physicist. After receiving her degree from MIT, she held postdoctoral fellowships at the Canadian Institute for Theoretical Astrophysics and Berkeley's Center for Particle Astrophysics and later fellowships with the Department of Applied Mathematics and Theoretical Physics at Cambridge and the Astrophysics Department at Oxford. Levin is now a professor of physics and astronomy at Barnard College of Columbia University. She is the author of *How the Universe Got Its Spots: Diary of a Finite Time in a Finite Space.*

It's Tuesday, and this is the number of times Einstein came to the forefront of my mind: eight. Eight times today I conjured up his name and image, no doubt a false image, a projection, a reflection of my own making—*my* Einstein.

(1)   6:25 a.m. This is no kind of hour to be awake. I am dragged up to consciousness under protest, on the decaying cusp of a dream. The room is as dark as midnight and warm as breath. The baby is standing with chubby fingers curled over the lip of his crib and eyes arched over knuckles, searching for me in the darkness. This isn't

179

where I expected to wake up. In the past few months, my husband and I have been shifting around the apartment in the small hours, looking for sleep, and we rarely remember our trajectories.

The disruption to my sleep (which could easily have lasted hours longer) elongated that beautiful window, just as I'm coming across from asleep to awake—when I am neither. As I cross over, a few important sparks from muddled dreams resolve into something meaningful, ideas that might survive scrutiny in the light of day. If I don't rehearse them quickly, they're gone. This morning I saw, as clearly as if I could hold it in my hand, how the shape of space might bend. There were twins traveling around the universe on clearly marked spirals, and I could almost resolve their orbits before the lines in my weak emulation of Einstein's dreams dissolved.

I am up out of bed and quickly forgetting the half-dream whose brightness dissipates in the dark room. I grab at the memory, but it crushes and deforms into nonsense under the direct assault. I scoop up the baby, drink in his sighs, and rest my head against his. The day begins.

(2) 8:25 a.m. I check my e-mail. John Brockman demands my essay on Einstein. "Cough it up!" he says. "I'm thinking!" is my reply.

(3) 8:50 a.m. Mark, a friend from London, calls, shouting at me: "I'm in Central Park! I'm in Central Park!" He's presenting a UK Channel 4 documentary on the neuroscience of genius, featuring a segment on Einstein's gray matter, and they're here to film. Ten minutes later, he turns onto Manhattan Avenue in South Harlem in a red convert-

ible and pulls up in front of my apartment, looking great, talking volubly about Einstein's brain, how he held it in his "grubby hands," how he had pieces of *Einstein's own brain* in his "shameful clutches." I'm mortified. So is Mark.

Mark and I spend the few minutes we have to see each other circling around on the sidewalk. We're punchy from suddenly being here, now, together in New York City, our voices raised above the buzz of the streets. We marvel at the baby, who shows off his stomping mobility, and feel time pass even as we hang relatively still, unable to stop it from whizzing by.

Mark and the convertible zoom away from my street and shrink out of sight. What does anyone want with Einstein's brain? Who is he to any of us? He lifted the spirits of generations bruised by world wars. He took our breath away with an epic demonstration of the grace of the human mind. His mind. His brain. A clump of matter, chemistry, and blood. Calcium, hydrogen, bone. Break his body into each fundamental particle and throw the bag of sand into space. Spare him the prodding dissection. That's not my Einstein.

(4)   10:05 a.m. Halfway through delivering the penultimate lecture for my Electricity and Magnetism course, trying hard to plow through Chapter 29 of the textbook, I sense that the circuit diagrams are no longer capturing my students' imaginations. I stop to remind them about James Clerk Maxwell and how he unified electricity and magnetism. How he discovered that light is a wave of electromagnetic fields. How he proved its speed is c. I want them to see the significance and the outright beauty of nature's

simplification of all the forces into a list of four that dictate the entire evolution of the universe, the emergence of life (including theirs), the spin of the galaxy on its axis. I tell them how scientists have labored to reduce the list of fundamental forces from four down to two, and how Einstein dreamed of an ultimate theory of unification, a reduction down to one—one law of physics, one magnificent natural proclamation from which it all derives, the whole universe and time and matter.

I tell them about Einstein.

My students are obsessed with their grades. Thirty-eight of the forty-two want to go to medical school. They are easily humiliated, and they are enraged by the difficulty of this subject—the math, the initial discomfort of the abstract concepts. A very few become petulant. The majority have risen up and impressed themselves and me. But they all need reassurance. "One of Einstein's professors called him a lazy dog," I tell them.

The story goes that Einstein liked to sleep ten hours a night—unless he was working very hard on an idea; then it was eleven. And while he slept the night and part of each day away, he dreamed. He dreamed of riding his bike through trees and catching the light as it fell off the leaves. He dreamed of time standing still as he traveled at the speed of light. He dreamed of relativity. He dreamed of curved space-time.

My own dream came back to me, wounded by my early-morning clumsiness and now brushed aside in favor of a circuit diagram with an alternating current, example 2 in Chapter 29. The class goes on.

(5) 1:05 p.m. Missed breakfast. Can't have lunch until I've dealt with breakfast. Very hungry but oddly distracted by a philosophy book on quantum mechanics and special relativity—*Quantum Mechanics and Experience,* by David Albert. I could get out of my office and forage for food, but hunger has made me too lazy for that. The book was given to me just yesterday by David himself, in his office in the Philosophy Department. I stare at the cover for a long time, then take up the book, leaning so far back in my chair that my arms and legs flail for a second to keep me from falling backward. Finding a reasonably safe equilibrium, not too far back, not too far forward, I start to read.

Einstein realized that space and time are relative. He came to this incredible conclusion through the idea that the speed of Maxwell's electromagnetic light is a fundamental constant. This simple assertion is actually outrageous. Our familiar experience is exactly contrary. For instance, if you race toward a moving train, its speed relative to you changes. But if you race toward a beam of light, its speed, as measured by you, will, remarkably, always be c. Einstein built a theory of space-time around the premise that the speed of light is constant and nothing can travel faster. In return, he gave up the idea that space and time were immutable absolutes. He invented relativity. The implications are surreal and Einstein worked them out in imaginative thought experiments. He envisioned an astronaut in a rocket on a round-trip, deep-space voyage at close to the speed of light while the astronaut's twin remained anchored on the ground. In order for the twins to measure the same speed of light, they must not experience the same passage of time. When they're reunited, they are no longer the same age: the astronaut is only a few years older; the earthbound twin is an old man.

Yesterday David Albert and I talked about this—the twin paradox. David marched to the blackboard in his office and drew two intersecting lines and talked about the inconsistency between quantum mechanics and special relativity. I marvel at how two simple lines can come to symbolize so many varied ideas: the scattering of two subatomic particles, Euclid's fifth postulate (which implies that parallel lines never intersect, and which turns out to be untrue on curved spaces), the X of tic-tac-toe. Today they symbolize the perfect relativity of two observers, the astronaut and his twin.

David talked about quantum mechanics' inconsistency with Einstein's special relativity. Quantum mechanics allows acausal events—events linked faster than the speed of light. And it may select one path through space as special, thereby violating the special relativistic premise that all measures are relative and no path is special. Which will win, quantum mechanics or special relativity? Which is right? He backed away from the cloudy blackboard, and we stared at the ambiguous chalky X. Frankly, it's confusing. But it's a bit stunning, too. I look forward to these meetings, here or at a coffee shop, when we can hammer at our various constructs to see which will buckle under the pressure and which will hold. Sometimes the physical ideas seem as sure and solid as a metal coin in the palm of my hand. But then the light changes and the coin turns into a wisp of plastic. We argue on.

(6)   3:15 p.m. I walk along Broadway. The sun is in my eyes, so penetrating that I can't see the street. On the periphery and high up, I see patches of things: the glare from a window above a storefront, the top of a pruned tree, a pigeon. Above that level, the white daylight is a blank canvas. I sink into the opacity of the hot sky. I imagine the universe is finite, closed like the surface of a sphere. The astronaut could orbit the earth forever, waving to his brother with every circuit. Each twin sees the other's clock run slowly. Neither is special. Time is relative. Which twin is younger? What time do they measure? Who observes what and when?

I reach into my pocket for a folded piece of scrap paper densely covered by calculations and diagrams and notes. Weak marks are written over with insistent forceful lines to remind me, when I read it again, of the answer. I unfold the crisp sheet and crinkle it into a shape somewhere near round. I crunch it in left hand and right, back and forth, enjoying the sound until the resistance of the paper weakens and it sounds soft, yielding like cotton. I lob it into the garbage, winding far behind my head, leaning back as though it were a solid lead sphere, or just a leatherbound ball, and hurtle it in a lovely arc as if casting off a massive weight. It flies freely, suspended and twisting, until an orange metal Dumpster rises with the turn of the earth to swallow it whole. The answer looked right. But it will look even better on a clean sheet of paper.

(7)   5:03 p.m. I am in string theorist Brian Greene's office. The window at the far end of the room is like a floodlight, so he appears in silhouette behind his big desk,

and this cracks me up. I fall into the chair to face him, and I feel diminished by its ridiculous proportions, the back easily reaching a foot and a half above me and the scalloped sides cradling me like a huge clamshell. This cracks him up.

I tell him about my conversation with the philosophers. We talk about twins and relativity and preferred slices through a finite space. It's a conversation run backward, starting with a conclusion and iteratively retreating a step at a time, finding some fault with each pace. Finally we move back to the proverbial square one. We take nothing for granted. Assume nothing. Believe nothing, except for the constancy of the speed of light. And from that one principle we make tentative steps forward again.

We do not have a clear question that we are trying to answer. It can take weeks for a well-posed question to emerge. A precisely stated objective is itself a huge sign of progress. For now, we're just testing what we know. Noodling about. Groping after intuition. We pick at a suggestion: Could there be a link among twins, finite spaces, and the consistency of quantum mechanics and special relativity? Will we build something inevitable from this collage or will it dissolve to mulch? The plot thickens.

(8) 12:35 a.m. I am listening to National Public Radio online, to a commentary on the first "astronettes" (female astronauts). As I'm listening, my eyes move in and out of focus. I'm staring at eight square feet of Plexiglas affixed to the blue-white wall in our front room. The corners have broken off where the nails went through, and rivulets of splinters emanate from weakened points, but from this close proximity it looks like a pristine surface. On it, I have

drawn intersecting lines, over and over again, the vertices highlighted with glowing blue or green or pink markers. A symbol of space. No limits.

I stare at the wall for hours. The baby's asleep, but there's a handprint near the bottom where he tried to contribute, in a last burst of enthusiasm for the day, before his mood collapsed in favor of sleep. I stare at the wall and try to visualize the path of a light beam as the astronaut tries to synchronize his clocks.

I am stressed-out and anxious, frustrated by the subtlety at the same time as it beguiles me. My temper is softened by the ridiculous image of Einstein and his big bulb of a nose, topping of wispy, impudent hair, overgrown mustache, shrouded smile—the image made all the more ridiculous when he closes his eyes and snoozes and dreams his way to truth. I feel the warm flood of sleepiness and the lure of my own dreams.

Then I have an idea for my Einstein essay for Brockman, and I start to write. He is there, in the front of my mind, *my* Einstein.

Einstein's Tie

MARCELO GLEISER

MARCELO GLEISER was born in Rio de Janeiro and received his undergraduate education there. He took his Ph.D. at the University of London in 1986. After a two-year postdoctoral fellowship at Fermi National Accelerator Laboratory in Illinois and a three-year postdoctoral fellowship at the Institute for Theoretical Physics of the University of California at Santa Barbara, he went to Dartmouth, where he is the Appleton Professor of Natural Philosophy in the Department of Physics and Astronomy. He is also the author of *The Prophet and the Astronomer: A Scientific Journey to the End of Time* and *The Dancing Universe: From Creation Myths to the Big Bang*.

Two weeks before my bar mitzvah, my step-grandmother knocked timidly on my bedroom door. As with most teenagers, my bedroom was a sacred fortress; not everyone in the family was allowed in. Reluctantly, I gave her the go-ahead. She was holding a flat package, carefully wrapped.

"Marcelo, I know how much you like nature, and how interested you are in science," she said. "So I thought of giving you this little gift. Here, take it." She handed me a package that reeked of decades of Brazilian humidity mixed with mothballs—a bit like my step-grandmother herself, Dona Ruth. I tore away the paper quickly. When I saw what it was, my jaw dropped: a photograph of Albert Einstein. Not only that, an *autographed* photograph of Albert Einstein!

"My brother-in-law, Isidoro Kohn, hosted him when he was in Rio in 1925," said Dona Ruth. "That's Isidoro, in the picture with Einstein. See the tie that Einstein is wearing in the photo? I still have it."

The tie! I looked at Dona Ruth with pleading eyes. She played tough. "You'll have to wait a bit for that one," she said, laughing. "Convince me you should have it." Good old Jewish bribery; never fails!

Dona Ruth died before she could give me the tie. It's with someone from her family, probably rotting in a closet. Not sure where. But the picture immediately went up on my wall. I made a shrine for it, with science and math books and a recording of Mozart's Double Violin Concerto in C Major flanking it. And I started to *really* study physics—beyond what was required in school. If Einstein could see into the invisible patterns that rule natural phenomena, I wanted to at least understand what it was that he had seen. In my teenage imagination, Einstein became a figure bigger than life: He loved music; he searched for a unified description of natural phenomena; he was Jewish, too—and a pacifist. I was bewitched by his Platonic view of the world, by geometrizing as a way of unveiling nature's deepest secrets. Johannes Kepler had done the same in the seventeenth century. The search for geometric patterns as the secret language of nature became an obsession with me. Still is.

I cannot deny that I am a bit of a rational mystic. But then, so was Einstein. And Kepler, although in a very different way. The rational Christian God became the rational in godless nature. So be it. It was when I found this out—that science addressed some of the most fundamental questions about existence—that I had the courage to take up physics as a lifetime occupation and not just an educational requirement. What a privilege it is to work for a goal larger than life, larger than time!

In an essay from 1930 entitled "What I Believe," Einstein wrote that "the fairest thing we can experience is the mysterious. It is the fundamental emotion which stands at the cradle of true art and science." No other quotation, by Einstein or anyone else, has made such a deep impression on me as this one. It beautifully synthesizes why I decided to become a scientist. Growing up, I first tried to find my answers in religion, as Einstein did. After all, my educators insisted that the biblical stories were true. And not just true, but the Sacred Truth, undeniable and unquestionable.

Death had come into my life very early, with the tragic loss of my mother. Reality was too hard and too painful. An impressionable boy, I looked for an alternative in the supernatural, the magical. Clearly, if the biblical stories were true, then there was an invisible reality parallel to ours, populated by angels and demons and promises of life everlasting. And if this alternative reality existed, I should gain access to it and reconnect with my mother. If she was in heaven, as I'd been told many times, she had to be an angel. And since she was an angel, she would surely come down to the mundane earth to visit her son. I couldn't accept the idea that she would simply dissolve into a fading memory. So I searched. And waited. And hoped.

Of course, apart from my developing a great fondness for gothic literature and horror movies, nothing happened. I grew disillusioned. I learned from history books how many people had died—had killed—in the name of religion, claiming God for themselves. Appalled, I realized that these crimes were still going on just as they had in the twelfth century. If faith was meant to inspire humanity to better itself, the project was clearly failing. By my early teens, my feelings had grown from disillusion to outrage. My belief in the supernatural, in the biblical stories, was shattered. There had to be another way to make sense of life.

Einstein's photograph came to the rescue. Here was a scientist—and not just any scientist but one of the greatest of all time—who claimed that the mysterious was a fuel for scientific creativity. But how could this be? Along with the biblical stories, I had been told that science was rigid, mathematical, grounded on the concrete; a scientist had to be the ultimate rationalist, a deductive machine. No room for feelings in science, and certainly no room for mystery. But I looked at the photograph and didn't see a deductive machine. I saw gentle, kind eyes; unkempt hair; a bushy mustache; someone whose concerns did not appear to be focused on immediate reality. I saw a scientist-artist, someone who blended intuition with logic, who fulfilled his spiritual quest for meaning through his scientific work. I saw science bringing human beings closer to nature. I saw a mentor.

Doubtless this image of Einstein is greatly romanticized. Quite possibly, had I the privilege of meeting him, I would have been disappointed in some way or other. Our icons are bigger than the real thing. They serve a selfish purpose: to give meaning to our lives. We deprive them of their own reality, making them (or what we think they represent) part of our expectations, our private universe. We need to adore someone or something, so that we can aspire to become bigger than ourselves. So I had my Einstein shrine, and his science became my religion, a godless religion accessible to all who cared to learn. And one filled with mystery.

Great scientists create when they are driven by a deep sense of purpose. In this they are no different from artists. At the risk of being called a generalist, I claim that the ultimate driving force behind all creative work is the awareness of our mortality. We create to remain in the present. Today we listen to Ludwig van Beethoven and John Lennon, we surround ourselves with pictures by Marc Chagall and Botticelli, we read Shakespeare and Virginia Woolf, we study Newton and Einstein. As we do so, their essence (or our perception of it) is still here, with us. As in Shakespeare's Sonnet 19, "Yet do thy worst, old Time! Despite thy wrong / My love shall in my verse ever live young."

Our immediate physical reality is Newtonian. Things that fall, fly, move, and flow; planets revolving around the sun (with the exception of a small correction in the orbit of Mercury); even the rotation of the Milky Way—all these phenomena are described by Newtonian physics. But as we stretch our perception into worlds that are virtually unobservable—the world of subatomic particles, of dying massive stars, of a 14-billion-year-old universal expansion—we dive into Einstein's reality. Newton unified the physics of the earth with the physics of the heavens, showing that the same set of laws described earthly and celestial motions. He did this by assuming a rigid structure for space and time: space as the stage whereupon things happen, time as the inexorable ticker of worldly events. Since this is precisely how we perceive space and time from our myopic viewpoint of slow velocities and human-sized distance scales, Newton's physics has a comfortable familiarity to it. It is a science that helps us to gain control over our sensory reality, make sense of what we see and measure. Not so with Einsteinian science.

Einstein's science was a raging iconoclasm, demolishing the very Newtonian notions of absolute space and time that were so cozy and nonthreatening. His science opened the door to an unknown world, one beyond sensory perception—an invisible world with mysterious properties and bizarre effects. Once you stepped into this new worldview, you couldn't go back. Like the mythic hero returning from his quest, you'd emerge transformed, with a new conception of reality. This was science as a rite of passage, science as spiritual fulfillment. Newton's ideas may well have had a similar impact on the minds of early-eighteenth-century natural philosophers, because they also revealed invisible connections between the heavens and the earth.

Gravity was seen as a mysterious invisible force, originating within matter by unknown mechanisms and propagating instantaneously through space with no obvious explanation. Still, Newton's science dealt with palpable reality, while Einstein's went beyond. A different icon for a different age.

No wonder Einstein became an obsession to so many. How could he not? In a world torn apart by the bloodiest war of all time, this Jewish scientist was proclaiming the existence of a reality wherein space and time are unified in a four-dimensional space-time, where space may contract and time may slow down, where matter is nothing but lumped-up energy. Who wouldn't want to step out of the miserable state that Europe was in in the early 1920s and into the rarefied atmosphere of a world beyond the senses? The same transformation gripped war-torn Athens around 400 B.C.E., when Plato led philosophy away from Socratic moral immediacy into an archetypal reality of forms and pure thought. Kepler did something of the same, although he was no icon during his lifetime, when the Thirty Years' War ravaged most of Central Europe: if there was no order down on Earth, there ought to be some up in the skies.

Einstein's signed photograph now hangs on my office wall at Dartmouth College. Well, actually not the original, since it would suffer badly from continuous exposure to solar UV radiation—I don't want his signature to fade any more than it already has in eight decades. During the past twenty years, apart from going a little deeper into Einstein's science and publishing dozens of papers based directly or indirectly on it, I have also learned more about his personal life and his moral dilemmas. It is hard for someone like me, who grew up in sunny Rio during the sixties and seventies, to fathom the excruciating hardship of living through two world wars, of being a Jew in Germany in the twenties and thirties, of witnessing the bombing of Hiroshima and Nagasaki and the development of the H-bomb. It is ironic that *Time* magazine, which selected Einstein as Person of the Century in its issue of January 3, 2000, had also pictured him on the cover of a July 1946 issue with a sad, pensive expression and in the background a mushroom cloud with $E = mc^2$ written on it. I wonder what went through the mind of this peace-loving man who used his celebrity to promote universal ideals of equality and democracy when he saw that cover. How quickly the patron saint of science was demonized as the scapegoat for a whole generation of physicists!

The development of the bomb, more than any other event in the history of science, redefined the role of the scientist in society. Since at least the time of Archimedes, whose military contraptions kept numerous divisions of the Roman army from taking Syracuse, science has been used for political ends. Today, as in the past, it is clear that the nation with the strongest military technology dominates the political arena. However, after the development of the nuclear bomb, this domination changed its face. For the first time in history, humankind had the power to annihilate itself many times over. Science turned apocalyptic. J. Robert Oppenheimer famously quoted the Bhagavad Gita after the first detonation at Alamogordo in July 1945: "Now I am become Death, the destroyer of worlds."

Although the nuclear-bomb technology and its related physics have very little to do with Einstein or his research, he has often been blamed for it. In countless movies and comics, the mad scientist bent on destroying the world has an Einstein hairdo and mustache and a thick German accent. While I was growing up, it was hard for me to reconcile the two images—scientist as creator and scientist as destroyer. I refused to see scientists as active players in the follies of lesser mortals. "No, they must have been manipulated by greedy politicians," I'd rationalize. But, with time, I came to realize that there is an inevitable connection between science and political power, that science serves both the best and the worst in humankind. Many of us take refuge in the rarefied atmosphere of academia, believing that this distance makes us immune to the dark side of what we do. But it doesn't. Science needs funding, and this funding, especially for basic science, comes from the government—a government that often adopts policies you may not fully support. To a greater or lesser extent, all scientists have dirty hands. Einstein knew that; he must have accepted his image on the cover of *Time* with the dignity of someone who had the moral duty to carry the heaviest burden of his age. Perhaps of all ages.

I look at the picture on the wall and realize that Einstein was my age when it was taken. His greatest contributions to physics were behind him by then. True, he still engaged in intense debates about the nature of quantum mechanics, in particular with Niels Bohr. But many of his colleagues, especially the young, viewed his resistance to accepting the intrinsically probabilistic properties of atoms as the philosophical prejudice of someone who was out of touch. Einstein knew he had a different mission ahead of him. If society turned him into an iconic figure, he could use this to further the causes he believed in. Deep down, all he wanted was to spend his days thinking about physics, immersing himself in the "mysterious," searching for the elusive unified field theory that would geometrize all of nature, the ultimate Platonic dream, but he also knew that withdrawing from the world would be immoral, that his voice had to be heard, that he could make a difference. In 1925, he went to South America to talk about relativity and raise money for the Zionist movement.

When Einstein arrived at Isidoro Kohn's home in Rio, the plan was to go straight to the hall where he was to lecture. Isidoro was taken aback when he saw that Einstein wasn't wearing a tie: the event would conclude with a formal dinner. "Professor Einstein, I must find a tie for you!" he said.

I can see Einstein rolling his eyes, weary of all the useless formalities he has had to endure. Why must people take themselves so seriously? "*Ach*, if you insist," he replies. Isidoro disappears into his bedroom and comes back with a dark, slim tie. "Here, you can borrow this one. Well, just keep it. It may come in handy."

Einstein puts on the tie. A photographer materializes, as if out of thin air.

"Professor Einstein, I hope you don't mind, but I'd be very grateful if I could have my picture taken with you," says Isidoro Kohn. "And if you could sign it afterward . . ."

"Of course!" Einstein replies, smiling.

Two days later, the package with the signed photograph arrives at Isidoro Kohn's home. To his surprise, it also contains the tie. A note, signed by Einstein, reads, "Mr. Kohn, thank you but I couldn't keep your tie. I want to have an excuse *not* to wear it any chance I have!"

I still don't have that tie. I'm hoping that by now I deserve it. Maybe Dona Ruth will intercede somehow, and I'll finally be able to complete my shrine.

The Greatest Discovery Einstein Didn't Make

ROCKY KOLB

EDWARD W. (ROCKY) KOLB is a founding head of the Astrophysics Group at the Fermi National Accelerator Laboratory, in Batavia, Illinois, and the director of its Particle Astrophysics Center. He is also a professor of astronomy and astrophysics at the University of Chicago. He received a Ph.D. in physics from the University of Texas and held postdoctoral appointments at Caltech and the Los Alamos National Laboratory, where he was the J. Robert Oppenheimer Research Fellow. Kolb has served on the editorial boards of *Astronomy* magazine and several international scientific journals. He was the recipient of the 2003 Oersted Medal of the American Association of Physics Teachers. He is the author of *Blind Watchers of the Sky*, which in 1996 received the Emme Award of the American Aeronautical Society, and coauthor (with Michael S. Turner) of *The Early Universe*, the standard textbook on particle physics and cosmology.

On a small planet orbiting just one of the billions of stars in just one of the billions of galaxies in the observable universe, a species evolved with the curiosity to look out into the outer reaches of space and into the inner recesses of matter and try to understand how it all works. Even so, it's really quite remarkable that anyone at all wonders about the origin and destiny of the universe. Most people put their feet on the floor every morning without thinking about their place in space and time. Perhaps that is a good thing. If your first thought of the day is the seeming insignificance of a single human life in the enormity of space and time, you might be tempted to pull the covers over your head and go back to sleep. But instead of being intimidated by the enormity of the universe, some people get out of bed and dedicate their lives to trying to understand its origin, evolution, and fate.

"I'm a cosmologist and proud of it!" That's the way I introduce myself to other scientists. It usually gets a laugh. Cosmology—the study of the origin, evolution, and fate of the universe—was considered a bit dodgy until relatively recently. It was once thought to be a disreputable science, just a bit more substantial than metaphysics, and certainly not "hard" science, whose ruthless experimental results restrain the imagination and speculation of theorists. Being a cosmologist was once nothing to be proud of. But cosmology today is not your parents' cosmology. A combination of precise observation and bold theoretical development has led to the construction of a standard cosmological model. Although this modern cosmological model involves the invention of new particles and interactions, it is firmly grounded in the standard model of particle physics. Our current cosmological model seems capable of describing the universe's origin from a singular event some 13.78 billion years ago, as well as its evolution from a formless, shapeless fog of elementary particles to the present cosmic structure of stars, galaxies, and galaxy clusters, and even suggests a new and unexpected destiny for the universe: an eternal, accelerating expansion.

Of course, no one really expects the standard cosmological model to be the final complete picture of the universe, but most cosmologists believe that it will be a big part of the final picture. Although a hundred years from now our cosmological model may well be criticized as naive in some aspects, it probably won't be found lacking in boldness and imagination. It addresses questions once thought to be forever beyond the scope of scientific inquiry. Einstein would have been proud!

The foundation of modern cosmology is his theory of gravity—general relativity. We cannot even address the basic questions of modern cosmology without it. Modern cosmology began shortly after its unveiling; in fact, the first paper on modern cosmology was written by Einstein himself, in 1917. That paper, "Cosmological Considerations on the General Theory of Relativity," was remarkable for several reasons.* It contains the concept of a cosmological constant, the outrageous idea that empty space has a mass-energy density—a sort of "weight of space." This curious idea was repudiated by Einstein himself in the 1930s, after the discovery of the expansion of the universe, only to be resurrected by cosmologists in the late 1990s to explain the

---

* *Sitzungsberichte der Preussischen Akad. d. Wiss.,* 142–52 (1917).

apparent acceleration of the expansion. But by far the most noteworthy aspect of the paper is what it fails to do. It should have predicted the universe from a Big Bang; this was the greatest discovery Einstein didn't make.

Einstein's 1917 paper on cosmology was written less than two years after his revolutionary paper on general relativity (the greatest discovery Einstein did make). More than anyone, Einstein must have realized that his theory of gravity opened new possibilities for understanding aspects of nature once thought beyond human comprehension. With his new theory of gravity in hand, he must have felt he had a powerful new weapon with which to confront the cosmos. For over a decade, he had pursued physics with seemingly unerring intuition, always on the mark, but when he turned his aim to cosmology he was far wide of the target.

When I teach modern cosmology to graduate students, I usually walk into the classroom on the first day of the term and without a word write a single equation on the blackboard. After silently admiring the equation just long enough for the students to begin to wonder whether they made a mistake in signing up for the course, I turn to them and say, "I assume you are familiar with the field equations of Einstein's theory of general relativity." I then proceed, in just a few quick minutes, to demonstrate that Einstein's theory of gravity does not describe a universe unchanging in time but rather one that is either expanding or contracting. Finally, I suggest that we will be particularly interested in the expanding solution, because that's the one that seems to describe the universe in which we live. *And the evening and the morning were the first day.*

When I look at Einstein's equations, "expansion" sort of screams out at me. Even skeptical students accept an expanding—or contracting—universe as an implication of Einstein's theory of gravity. But for over a decade after Einstein developed his theory, he could not hear what his own equations were saying. I have often wondered how Einstein missed this one. How did Einstein miss the opportunity to predict the expansion of the universe?

Perhaps when describing how naturally the expanding universe follows from Einstein's theory, I am playing the role of the armchair cosmologist, who can see all the pieces of the problem neatly laid out on the board and how the game plays out. At the frontier of any science, things never appear quite as clear as they later do. Now it is clear that Einstein should not have introduced the "slight modification" to his equations for the purpose of keeping the universe from expanding. He should instead have predicted that the universe expands! Ironically, we now believe that his slight modification was correct, but for the wrong reason: the modification does not prevent the universe from expanding but causes space itself to expand at an ever-increasing velocity.

In 1905 Einstein looked at a jumble of equations on the papers spread across his desk in the Bern patent office and realized that they told him that space and time were not independent but interrelated in a deep way. Somehow, he was able to reject the assumed independence of space and time, which had been a part of our worldview for longer than history can chronicle, and imagine a unified space-time. Ten years after that leap of imagination, after a great struggle, he realized that his equations told him that space needn't be flat but would curve in the presence of matter. He was able to dismiss the flat-space geometry of Euclid that had ruled mathematics, physics, and cosmology for two millennia and embrace the curvature of space.

Then, two years after his 1915 discovery, he learned that the theory he had spent years and almost unimaginable intellectual effort developing did not admit solutions describing the universe as he sensed it must be. And so in 1917, the man who had changed his perception of space and time because his equations told him to, who had altered his picture of the geometric construction of space because his equations were telling him to, could not hear what his equations were telling him about the expansion of the universe. Rather than predict the expansion, he concluded that his beautiful equations of general relativity "need a slight modification."

How did Einstein miss the mark? In his 1917 paper, he was looking for a cosmological model that had two properties:

1) The universe is unchanging in time—that is, it is *static*;

2) Space in the universe does not exist without the presence of matter.

Newton's theory of gravity did not satisfy either of these conditions, and Einstein thought his grander edifice might.

The first condition is just a prejudice that reflected Einstein's view of the cosmos. Whether he believed that the universe had existed forever or was created in a special event (say, requiring six days of effort followed by a day of rest), he assumed that it had always had more or less the same appearance it did now. He knew that things move around: the planets orbit the sun, and the stars exhibit small motions through our galaxy. But he thought that all these motions were insignificant and that the universe today looks pretty much as it always did. This was the generally held scientific view of the universe in 1917. When he began his cosmological investigations, Einstein did not expect the universe to expand. (Of course, when he began his investigations into relativity, he did not expect space and time to unite and space to curve.)

In fact, few astronomers at the time grasped the true nature of the universe. The standard cosmological model of 1917 held that there was but one galaxy in the universe, our own Milky Way. Not until 1924 did the American astronomer Edwin Hubble prove the existence of galaxies distinct from ours. Five years later, in 1929, Hubble demonstrated the expansion of the universe, the greatest discovery Einstein did not make.

Einstein had earlier consulted astronomers about the possibility of detecting the bending of light as it passes close to the sun, but there is no record that he asked astronomers about the possibility of the universe expanding. Had he done so, he might have learned the peculiar fact that "spiral nebulae" (which turned out to be those other galaxies) seemed to have an anomalously large velocity of recession.

The second condition Einstein demanded of his cosmological model was a reaction to the concepts of absolute space and absolute time, which are at the foundations of Newton's dynamics. Newton writes in the Scholium to

Book I of *Principia Mathematica,* "Absolute, true, and mathematical time, of itself, and from its own nature, flows equably without relation to anything external. . . . Absolute space, in its own nature, without relation to anything external, remains always similar and immovable."

In Newton's world, stars and galaxies are just players on the fixed field of space and time, moving through the universe according to the laws of Newtonian physics. But in Einstein's universe, space and time participate in the game as well and, in the interplay among space, time, and matter, space and time are not absolute but relative.

Einstein's rejection of absolute space and time seemed to arise from two sources: the first was, of course, his discovery of their relativity; the second was the influence of the physicist and philosopher Ernst Mach. Mach proposed that the inertial forces acting on an object come from all the stars and galaxies in the universe. To Mach, it was meaningless to talk about the motion or rotation of a solitary object in an otherwise empty universe. Einstein himself referred to this idea as "Mach's principle." In his later years, he repudiated Mach's principle, but in 1917 it influenced his work on cosmology and contributed to his belief that space would not exist without matter.

Einstein initially tried to find a cosmology describing a universe infinite in space that satisfied his desiderata, but he quickly realized that such a solution could not exist. Then he tried to find the solution in a universe of finite spatial extent with the same two conditions—and failed again. He discovered what we all know now: that a universe containing matter and radiation just doesn't want to stay put but naturally seems to want either to expand or to contract. But instead of questioning the original goals he had set for his cosmological model in light of what he had found, Einstein chose to butcher the general theory of relativity, a theory he had spent so much time and effort perfecting. In his 1917 paper on cosmology, he wrote:

> I shall conduct the reader over the road that I have myself traveled, because otherwise I cannot hope that he will take much interest in the result at the end of the journey.
>
> The conclusion I shall arrive at is that the field equations of gravitation which I have championed hitherto still need a slight modification. . . .

The "slight modification" consisted of adding to his beautiful equations an ugly new term whose sole purpose was to keep the universe static. The addition is known as the cosmological term, or the cosmological constant, and it can be adjusted in such a way as to balance the effect of the curvature of space that tends to drive expansion.

This paper of Einstein's has caused more than its share of puzzlement among physicists and historians. Up until then, Einstein's instincts as a physicist had rarely led him astray; his intuition and insight had been uncanny. Perhaps it is too much to expect anyone to have predicted, in 1917, the expansion of the universe, and yet Einstein was not just anyone. Perhaps his ingrained sense of how the universe should be was deeper than his sense of space, time, or geometry.

As a cosmologist, I take away a few lessons from Einstein's first foray into cosmology:

It's OK to be wrong. Even Einstein was wrong in his cosmological work. Of course there is a difference between being wrong and being stupid.

Listen to what the equations are telling you. Sometimes the tune is subtle: no less a maestro than Einstein was deaf in this case.

Simpler is usually better. The expanding universe—in my opinion, at least—is a far simpler solution than Einstein's 1917 cosmological construction.

Don't be afraid to make bold predictions. Perhaps Einstein was too conservative when he failed to predict the expansion of the universe.

Be unprincipled. Principles are dangerous things. Einstein was led astray by Mach's principle. My friend and fellow cosmologist Andrei Linde always says that even people without ideas can have principles.

Sometimes you can be right for the wrong reasons. Einstein introduced the cosmological term to keep the universe static, which it is not; but recent observations suggest that the cosmological term is present and causes the expansion to accelerate.

Admit when you are wrong. After Edwin Hubble discovered the expansion of the universe, Einstein readily described the cosmological term as a blunder.

Never admit when you are wrong. The usual joke among cosmologists today is that if Einstein had not admitted he was wrong and had instead stuck to his original idea of a cosmological constant, he could have been famous.

Perhaps the most important lesson is to *try*. My Einstein was not intimidated by the size and complexity of the universe but summoned the courage and imagination to attempt to elicit its mysteries, even if its expansion was the greatest discovery he never made.

The Gift of Time

RICHARD A. MULLER

RICHARD A. MULLER is a professor in the Physics Department of the University of California at Berkeley and faculty senior scientist at the Lawrence Berkeley Laboratory, where he is also associated with the Institute for Nuclear and Particle Astrophysics. He received his Ph.D. in elementary particle physics but subsequently moved into astrophysics and geophysics. His research interests include the anisotropy of the cosmic microwave background, supernovas, the origin of the earth's magnetic flips, impact catastrophes, and glacial cycles. He is the author of *Nemesis: The Death Star* and (with Gordon J. MacDonald) *Ice Ages and Astronomical Causes*.

He who steals my purse, steals trash. He who takes my time, steals my life. Einstein left me many gifts. He set up the framework for the discovery that led to my faculty appointment and tenure: the cosine variation of the cosmic microwave background radiation. His equations of gravity led to a system for understanding the universe as a whole, and that inspired my work to set up a search for supernovas to measure what I thought would be cosmic deceleration. My former student Saul Perlmutter completed this project and discovered (simultaneously with another group) cosmic acceleration. Much has been made of Einstein's regrets over including—or not including—the cosmological constant in his equations, but as a good physicist and mathematician he had no choice, and this constant accommodates (without explaining) the strange acceleration. Yes, I have much to thank Einstein for. But more than anything else, I thank him for the gift of time.

Time is a remarkably elusive concept. Some treat it as a mere coordinate, a way to help specify an event. If you do so with three spatial coordinates (x, y, z), then time becomes the "fourth dimension"—but only in a trivial sense. In this manner, it appears in most physics equations. But although physics uses time, it is our dirty little secret that we don't really understand time. Physicists will tell you that time is now "unified with space" (thanks largely to Einstein), and we are supposed to be happy with that. But time behaves in a fundamentally different manner from space, in a way that physics doesn't quite acknowledge. Time is significantly more mysterious than space.

Consider, for example, the "flow" of time. We are all aware of the fact that time appears to be moving—but if we try to define what we mean by this, we get hopelessly lost in a tangle of circular definitions. Time is not actually going anywhere, but it does appear to be progressing. Our sense of the progress depends on our biology; a fly can respond much faster than a human, because our sense of time is determined in part by our ability to sense and think and move and remember. Roughly speaking, we might say that our sense of time is set by our heartbeat, although the round-trip time from a nerve tip to the brain might be a better quantization.

But in quantifying the motion of time, have we really made any progress in understanding it? I don't think so. We often fall into the trap of naming things we don't understand, thereby giving the illusion of the advance of knowledge. The rate of time, measured by bodily function, does not appear to be constant. Recovering from an illness, or attending a boring lecture, can seem to stretch time, whereas fun seems to compress it. To avoid the nonsynchrony of the human time sense, we build clocks based on physics, devices that don't get as emotionally involved with the phenomena around them. That's why we use clocks to measure the pace of events. When Einstein was a patent clerk, one of the pressing problems of the day was how to synchronize clocks at different train stations. There is a sweet irony here, because when he ultimately developed his theories of relativity he showed that such synchronization is in principle fundamentally impossible (although it can be done well enough for the purposes of train travel).

But why is there a "pace of events"? Space doesn't move forward; why does time? You can go backward and forward in location but not in time. Time seems to be really different. What is going on?

Maybe you've wondered about these questions. If you are not a physicist yourself, you may have assumed that the answers lie in the equations of physics. Let me assure you that they do not.

A good way to avoid the question is to replace it with

one that we can at least address. Instead of explaining the motion of time, let's see if we can account for its direction. Just look up on the Web, or on Amazon.com, the "arrow of time." There have been countless articles and numerous books devoted to this subject, and for the most part they are full of interesting physics and important insights.

The "arrow of time" refers to the fact that time appears to progress in a definite direction. The question has great substance. Here is a startling way to approach it: Why do we remember the past rather than the future? That seems like a silly question, until you recognize that in a completely mechanical, deterministic universe the future of all atoms is simply a function of their past. Information doesn't change with time, only the way it is organized.

But we don't live in a mechanical universe. Quantum mechanics, with its probabilistic ways, clearly needs to distinguish past from future. Probability relates the future to the past. Current (please let me use that word) theories assume that the past determines the future. Since we can't explain why, we give that determination a name: causality. But causality is not a principle of physics; it is an assumption that may not be true, a name given to something we don't understand. The mystery of causality is linked to the mystery of time's arrow. Most physicists will feign understanding of time. They'll tell you that its direction is set by the increase of entropy, and they will link this to another unexplained fact: that in the past, the universe was much more highly ordered than it is now. It was unlikely to remain that way, and so as we move to a more probable state, we experience time.

"Entropy" is physicist jargon for a number that represents the probability of a given situation. (Technically, it is the logarithm of that probability, but that's not important for our purposes.) As the universe evolves into a more likely state, the entropy increases. And that is what determines the arrow of time. Consider two photographs of the universe. See which one has the more confused arrangement of atoms and photons. That is the one taken at the "later" time.

The whole issue is actually a little more subtle. If we are to remember the past (rather than the future), then the entropy of our brains must actually decrease; that is, the connections of our nerve cells must become more organized. We can do that—local entropy doesn't have to increase, only global entropy. So in the process of learning, we throw off heat, some of it in the form of infrared radiation, and most of it eventually turning into such radiation, which is then expelled to space. If we include this radiation, then entropy is increased—as it must be—but meanwhile the entropy of our brain is decreasing. Memory represents decreased entropy, and it is memory that determines the arrow of time!

It isn't only the entropy of the brain that decreases. The entropy of the sun is decreasing. So is the entropy of Earth. The entropy of the atoms in the universe is actually decreasing! It is only when you include the entropy of their emitted radiation that the entropy of the universe is going up.

Virtually all of the entropy of the universe is in photons (mostly microwave radiation) and perhaps neutrinos.

More than 99.99999 percent of the entropy of the universe is already contained in radiation, and the entropy of that doesn't change as the universe expands. Put this way, we can say that the arrow of time is determined by the tiny little bit—the 0.00001 percent—of the entropy that is decreasing. Can you really believe that it is this tiny component of entropy that gives us memory and determines the arrow of time?

Confusing, isn't it?

To confuse you, and distract you in the process, was my goal. By now you have probably completely lost sight of the fact that I haven't addressed the question of why time moves at all. I have used the politician's trick of tergiversation. Physicists, largely thanks to Einstein, claim to have "unified" space and time. Time is no longer the fourth dimension in a trivial sense but part of a four-dimensional space-time system that can be rotated, stretched, spun, even twisted. Yet time moves and space doesn't. Why?

Let me try to confuse you a little more. Physicists like to think that there is nothing sacred about three spatial dimensions. Some of the modern theories of elementary particles posit that there are ten spatial dimensions and that seven of them are rolled up into small regions, like the surfaces of strings, so that in our large, macroscopic world we are not aware of their existence. There are only three extended spatial dimensions left. Although it's a bit abstract, I will assume you have no difficulty with that. But imagine for a moment a theory that has two time dimensions. Mathematically, that's easy; we can even work out calculations. But what does it mean for our perceptions?

If there are two time dimensions, would both of them "progress"? Would both times move?

There is no problem with the physics, because physics does not address the issue of the motion of time. Physics thus deftly avoids the problem. If you believe, as many of my colleagues do, that something that cannot be described by physics is something that has no meaning, then we have reached the end of the discussion. *Whew!* We have defined the question away.

But I don't accept that. Thanks largely to Einstein and the progress he made in understanding time, I think there's hope of answering it. Present-day physics doesn't understand time, but Einstein showed us that we should not give up.

Here's another obvious concept that has yet to be incorporated into physics: the deceptively simple idea of "now." Read this word now. You just did. But ponder the word you read. You read it a second or two ago. What does that word mean? Does it mean what it did when you read it?

The concept of "now" does not exist in physics. Physicists can cover this failing and even make it sound like a positive achievement: "The laws of physics are invariant under time translation." That is true—at least for the laws of physics as they currently exist in the textbooks. That means that "now" has no significance, no meaning. But you know what I mean by the word, don't you? Is something missing from physics?

Is my now the same as your now? Remarkably, Einstein gave us enormous insight on this subject. He was able to show that the concept of simultaneity could not be put on a universal basis. That means there is no universal now, good in all reference frames. Two events that are both occurring now according to a clock in reference frame 1 would not necessarily be simultaneous in reference frame 2. Philosophers can debate and discuss the meaning of now in long essays, but anything they say will be rendered meaningless if it doesn't take into account the Einstein result.

That does not mean that "now" has no meaning. You know what it means to you, even if your now is different from mine. There is still something there that physics has not yet addressed; some people think it never will. But Einstein's equations show us that the concept of time is subject to a nonsubjective description.

Just as startling as the loss of simultaneity is the conclusion that Einstein reached concerning time intervals. Take two people, John and Mary, twins who get together to celebrate their twenty-first birthday. Then they go off and travel. They both experience time: they both age. They get back together when John is celebrating his twenty-second birthday. But unless their travels matched precisely, they will not both have experienced the same amount of time. If, for example, John stayed at home and Mary spent the year in a circular path moving at 80 percent of the speed of light, then when she returns she will not be a year older. She will have experienced not a year of life but only a little over seven months.

This is a hard and well-established law of physics. For a moving object, time slows down by a factor of the square root of $1 - b^2$, where b is the velocity expressed as a fraction of the speed of light. This was, in my mind, Einstein's greatest achievement. The factor had appeared before, in the transformation devised by Hendrik Lorentz, but many people give Einstein the credit for recognizing that it was not a fluke, a funny property of the behavior of (possibly incomplete) Maxwell's equations, but a valid description of the way time behaves. Time intervals are plastic. The time between two events (in this case, the two birthday parties of John) depends on the path you take between them. That is a fantastic result.

Einstein, in taking the weird behavior of time seriously, then applied it to mechanics. The result was his most famous equation, $E = mc^2$.

Even if it isn't really required for train timetables, the rubbery quality of time, sharply defined by Einstein's equations, is an easily verified result—not when Einstein first proposed it, but today in a modern laboratory. When I was working on my Ph.D. thesis, I verified it nearly every day. I could see the lifetimes of elementary particles extended whenever they were traveling fast, and they followed the Einstein/Lorentz equation. At Berkeley, we have an undergraduate laboratory in which students study the behavior of cosmic-ray muons; they can see the time-dilation effect themselves. Fast muons experience less time and therefore live longer.

This behavior of time is utterly wonderful. It is tricky to understand, but if you spend enough time teaching it (lucky me), then it all begins to seem natural. The twin paradox just described is not a paradox at all but an "effect."

The real questions have to do with the movement of

time and the nature of now. Physics has made no real progress on these conundrums. Most theorists aren't even trying. Instead, they are taking on more straightforward issues, such as dimensionality. But Einstein's progress showed us that we have hope. Time is at least partly amenable to physical analysis. Maybe "now" and the flow of time will be forever hidden, never explicable to the equations of physics—maybe even remaining outside the realm of physics, in the world of the spirit. But it is too early to abandon our attempts. Maybe lurking out there, somewhere in the world of physics and mathematics, is a new set of equations that will stretch our minds and account for these mysteries.

Accountants love problems where all the pieces fall into place. Physicists love confusion, mystery, and being surrounded by things they don't understand. Prior to the theory of relativity, time was just a coordinate—something you explained other things in terms of. After relativity, it has become something with its own behavior, and we have made only a little progress in understanding it. Einstein showed us that at least some aspects of time are amenable to analysis. He gave me, all of us, the gift of time.

Flying Apart

PAUL C. W. DAVIES

PAUL C. W. DAVIES is a professor of natural philosophy in the Australian Centre for Astrobiology at Macquarie University, Sydney. He previously held academic appointments in astronomy, physics, and mathematics at the Universities of Cambridge, London, Newcastle upon Tyne, and Adelaide. His research has spanned the fields of cosmology, gravitation, and quantum field theory, with particular emphasis on black holes and the origin of the universe. He is the author of, among other books, *The Mind of God, Other Worlds, God and the New Physics, The Edge of Infinity, The Cosmic Blueprint, Are We Alone?, The Fifth Miracle, The Last Three Minutes, About Time,* and *How to Build a Time Machine.* His awards include the 2002 Faraday Prize of the Royal Society and, for his contributions to the deeper implications of science, the 1995 Templeton Prize.

Albert Einstein may have been a genius, but he was not immune to making mistakes—or, at least, errors of judgment. He refused to accept quantum mechanics, a subject he helped create in his annus mirabilis. "God does not play dice with the universe," he famously quipped. Events have proved Einstein wrong: experiments with atoms and photons have demonstrated beyond doubt that quantum uncertainty is an intrinsic and unavoidable feature of the physical world.

Ironically, the proposal that Einstein himself regarded as his "greatest blunder" might have been right all along. This was a late modification—sometimes unkindly called a fudge factor—that he made to the crowning achievement of his career, the general theory of relativity. I became fascinated with Einstein's fudge factor when I was a student in the 1960s. Unfashionably, I found it tantalizing rather than repugnant, and over the years I have argued in its favor in the face of widespread contempt for it. Now the tables are turning, and scientists are reluctantly admitting that maybe Einstein was wrong to think he was wrong.

The issue has to do with the nature of gravitation and the expansion of the universe, and it epitomizes Einstein's distinctive style of thinking. The young Einstein didn't know that the universe was expanding. Nobody did, until the painstaking observations of Vesto Slipher and Edwin Hubble in the 1920s demonstrated that the galaxies are flying apart.

Einstein's special theory of relativity, published in 1905, was a theory of space, time, and motion. It made no specific reference to gravity, but it heralded the end of Isaac Newton's theory of gravitation, for a very simple reason. According to Newton, gravity acts instantaneously over a distance—that is, if the sun wobbles, the earth will wobble in its orbit at the very same moment. But Einstein's theory of relativity requires that no physical effect can propagate faster than the speed of light, and light takes some eight minutes to cover the 93 million miles between our planet and the sun.

Replacing Newton's theory of gravitation cost Einstein another ten years of hard work. The upshot was his general theory of relativity, published in 1915 and sometimes touted as the greatest intellectual achievement in history. It certainly was—and is—a magnificently elegant and powerful theory. It describes gravitation not as a force at all but as a warping, or distortion, of space and time (strictly, of a unified space-time), although the theory can be recast in the language of forces when necessary.

The central problem Einstein faced was to write down a set of equations describing the gravitational field created by a given distribution of matter. He was guided by a deep philosophical conviction that the theory should be geometrical—gravity as warped geometry—but not in a way that depends on the viewpoint of the observer or the coordinates the observer uses to make measurements. An analogy would be to say that the curvature of the earth in the vicinity of some point P should not depend on whether P's longitude is measured from the Greenwich meridian or from a meridian running through, say, Mecca.

All these restrictions hemmed in the possible mathematical structures Einstein could employ to describe the gravitational field. Even so, there are still a countless number of possibilities. Einstein did what physicists often do when there is a bewildering choice: he kept it simple. It turns out that the simplest term describes a force quite unlike familiar gravitation. For example, it increases with distance, whereas normal gravitation diminishes. So Einstein discarded that and looked at the next simplest. Here he found a close match to normal gravitation. In fact, he was able to show that the second alternative mimics Newton's description of gravity—that is, as an inverse-square-law force—whenever the gravitational field is weak and the attracting bodies are moving slowly compared with light.

This was a great triumph, and the theory was hailed as a spectacular success. But one feature of it baffled Einstein. Gravitation has always been a subject of importance in astronomy and cosmology. On the scale of billions of light-years, gravity dominates over all other forces. So a decent theory of gravitation needs to give a plausible description of the universe as a whole. Einstein set out to construct a cosmological model, making the simplifying assumption that the matter in the universe is spread uniformly—as indeed it is, on a very large scale.

The puzzle Einstein encountered is easily described. If gravitation is always attractive, then why don't all the stars simply fall together in a big crunch? Newton had likewise wondered why the universe didn't collapse under its own universally attractive gravity, but he managed to sidestep the issue. Einstein confronted it head-on and came up with a ready-made solution. The first term of his field equations—the simplest term, which he had initially discarded—describes a force that can be either attractive or repulsive. The strength of this force is undetermined; it is an arbitrary parameter in the theory. Einstein reasoned that by picking the right strength and making it repulsive, this force could be tuned so as to exactly cancel the attraction among all the stars. In effect, the repulsion would shore up the cosmos against its own immense weight.

The repulsive force can be thought of as a type of antigravity. Because it grows in strength with distance, its effects are negligible on the scale of, say, the solar system—which is just as well, because the original theory already gave excellent results there. But on a truly cosmic scale, a uniform static universe is possible, exactly the way most astronomers of the day assumed it should be. Einstein duly published his static model of the universe in 1917. The addition to the field equations is known as the cosmological term, and the numerical parameter it contains—the factor that determines the strength of the repulsive force—as the cosmological constant.

This was not Einstein at his best. For a start, the cosmological term did indeed have an air of fudge about it. There was no other reason to change the original theory, and scientists dislike ad hoc solutions to problems. Second, it is obvious that the cosmic balancing act between attraction and repulsion is unstable. If the universe were to shrink a bit, then normal (attractive) gravity would get slightly stronger but the cosmic repulsion would become weaker. This would upset the delicate balance and cause the universe to shrink further, culminating in the very collapse Einstein sought to avoid. On the other hand, if the universe were to grow a bit bigger, antigravity would gain the upper hand and runaway expansion would ensue.

What did it for Einstein, however, was Hubble's discovery, a decade or so later, that the universe isn't static anyway, but expanding. Had Einstein stuck with his original version of the general theory of relativity, he would have been forced to conclude that a static universe is impossible and would almost certainly have predicted Hubble's results. Einstein's chagrin at having missed such a momentous opportunity led him to abandon the cosmological term in disgust, describing it as the biggest blunder of his life. Not surprisingly, cosmic antigravity became decidedly unfashionable among astronomers thereafter.

And that is pretty much the way things stood when I was a student in the 1960s. But I have always been fascinated by things that are possible yet out of favor. Just because cosmic repulsion wasn't needed to make the universe static doesn't logically imply that it is nonexistent. After all, this is the *simplest* term in a possible equation for the gravitational field. It has an adjustable parameter multiplying it—the cosmological constant. After his U-turn, Einstein claimed that he knew the value of this adjustable parameter and that it is zero. But how can he have known? The whole point about adjustable parameters is that they have to be measured by observations, unless some deep principle can be invoked to argue that a particular value is uniquely necessary. But no such principle was forthcoming, only prejudice.

How could we tell if the cosmological constant is nonzero? Well, it would show up in the way the universe expands, which would be a bit faster than might otherwise be the case. Until recently, astronomers couldn't even agree on how fast the rate of expansion is, let alone determine any small deviations in it, so the observations were consistent with a nonzero value but did not compel it.

To my mind, there needed to be some other reason why a nonzero cosmological constant was a good idea— something from a different branch of science. In the 1970s I found another reason. I was working at that time on the quantum theory of fields, a topic seemingly far removed from cosmology. A perennial preoccupation of quantum field theorists concerns the properties of the vacuum— that is, empty space. According to quantum theory, total emptiness is a myth. To be sure, one can remove all the atoms, all the photons, and so on from some region of space. But there remains a residue that cannot be eliminated. This residue consists of so-called virtual particles. They exist only fleetingly, springing into being spontaneously from nowhere and disappearing again in pretty short order. These comings and goings are driven by quantum uncertainty—the very thing Einstein hated. Random surges in energy over very short intervals of time create virtual particles from nothing. Fluctuations the other way destroy them. Thus virtual particles flicker in and out of reality, enjoying a sort of half-existence.

These ghostly entities mostly go unnoticed, but they can leave distinctive imprints in the properties of atoms and other quantum systems, so we know they are there. The question I and other theoretical physicists were asking in the 1970s was whether or not these virtual particles gravitate in the same way as real particles. Does all this ephemeral quantum stuff have real pulling power? And if so, how much do virtual particles weigh?

Calculations showed that you can't separate the real stuff from the virtual stuff in the gravity game: it all contributes. But the startling result was the manner in which the virtual particles gravitated. The precise details depended on the geometry of the background space-time in which the virtual particles made their brief appearance, but there was always a contribution that exactly mimicked Einstein's cosmological term. Thus quantum vacuum effects automatically generate cosmic repulsion. So it isn't so much a question of why an antigravity term should be included in the gravitational field equations as why it should be left out.

The burning question, of course, was *how much* antigravity the quantum vacuum generates. Here we ran into trouble. Taken at face value, the calculations suggested that the total amount of energy swarming about in empty space is infinite, leading to an infinitely strong antigravity force. This was an embarrassment of riches!

Whenever infinity crops up in a calculation of an observable physical quantity, alarm bells ring. Obviously an unjustified assumption had been made in these calculations. It wasn't hard to see where the trouble lay. The contributions to the cosmic repulsion made by the virtual particles depend on their duration. The ones that rapidly flicker in and out of existence exert a bigger (anti)gravitational effect than those that last longer. Our calculations assumed that there was no lower limit to the time a virtual particle might exist. But this was doublethink. The same quantum fluctuations that make the virtual particles also afflict the space-time in which they live, causing it to jiggle about. It's a tiny effect, but on a truly microscopic scale of size and duration these quantum fluctuations could cause drastic modifications to the structure of space and time. That scale had been known for many decades and is named after Max Planck, the originator of quantum theory. The Planck time is about 10^{-43} seconds. Physics folklore says you are not justified in dividing time into intervals shorter than that.

The Planck time thus provides a natural lower limit on the lifetime of a virtual particle. Inserting that limit into the calculations for the antigravity effect produces a finite result, but it is still a headache. The strength of the repulsive force works out to about 10^{120} times the maximum value permitted by observations. That is too large by a factor of one followed by 120 zeros!

This spectacular mismatch of theory and observation became known as the cosmological-constant problem. For a couple of decades, many physicists thought the problem would be solved by going to a more sophisticated theory, perhaps one in which some virtual particles contributed negatively to the cosmic repulsion, exactly canceling the positive contributions, to make zero after all. Einstein would have liked that. However, attempts along these lines were unconvincing, and it seemed to me that the old prejudice against the cosmological term was hard at work. Obviously there must be some suppression mechanism that drastically reduces the strength of the cosmic repulsion, but there is no sound reason why it has to reduce it to precisely zero.

Let me summarize the position as I saw it in the early 1990s. The most general gravitational field equation consists of a succession of possible terms, each multiplied by a strength parameter that must be determined by experiment or observation. The simplest term is the cosmic repulsion, the next looks like Newtonian gravitation. There are more complicated terms, too. Einstein finally decided to go with the second term only, and fix the strength parameters of all the others at precisely zero. But without a theory of these strength parameters, picking zero is completely unjustified, especially as simple-minded quantum vacuum calculations give (very large) nonzero values. Or so I argued. I even wrote a book (*About Time*, published in 1995) sticking my neck out in favor of a nonzero cosmological constant. Very few physicists or cosmologists were prepared to go along with my line of argument, however—a notable exception being George Efstathiou, then at Oxford University. Einstein's legacy was so strong, and his antipathy to the cosmological term so legendary, that only direct observational evidence for cosmic repulsion would change sentiment.

Then, out of the blue, that very evidence was discovered. In the late 1990s two international teams of astronomers announced that the universe was expanding faster than it should be if its expansion was subject to attractive gravity alone. Their conclusion was based on the study of distant supernovas. These are violent events that afflict certain compact stars and can be used to work out the stars' distances, which in turn give a measure of how the rate of cosmic expansion varies with time.

Let me explain the conventional picture. The standard model of the universe starts with a Big Bang, now known to have occurred 13.7 billion years ago. The universe began expanding exceedingly rapidly. But the gravitational attraction of all the matter served as a brake on the expansion, progressively slowing its rate in the same way that a ball thrown in the air slows as it climbs. For a given density of cosmic matter, this expected deceleration could be calculated and compared with observation.

Consider now how the expansion is affected if a cosmic repulsion force is included. Recall that this force is small at short distances. Just after the Big Bang, when the universe was very compressed, the repulsion would have been too weak to make much difference, but as the universe grew larger, the repulsive force would increase, until eventually it would prevail over the attraction. At this point, the expansion of the universe would cease decelerating and start accelerating. This is pretty much what astronomers claim to have observed, with the turnaround occurring about six billion years ago. Einstein would doubtless have been nonplussed were he alive today.

Though there are several theories that might account for the accelerating expansion, the simplest explanation remains Einstein's cosmological term, or, equivalently, the energy of the quantum vacuum—today dubbed "dark energy." But the matter cannot be laid to rest here. The original cosmological-constant problem persists. We still need to explain why the amount of dark energy is so much smaller (120 powers of ten, remember) than the natural value suggested by quantum field theory.

Even the most ambitious attempts to produce a unified theory of physics, such as the currently fashionable string theory or the related M-theory, give little purchase on this problem. In fact, some leading theorists have suggested, perhaps with a hint of desperation, that the strength of the cosmic repulsion is a random variable, its tiny value in our region of the universe being just a fluke. Take a God's-eye view, they claim, and almost everywhere the repulsion will be vastly stronger. In typical regions, the cosmic repulsion would be so strong that all the matter would explode apart before any stars or galaxies had a chance to form. The reason we find ourselves living in such an atypical cosmic location, so the argument goes, is that life would be impossible in the explosively expanding regions. As a cosmic real estate agent might put it, the key to life is location, location, location.

I am quite sure that Einstein would have hated this anthropic explanation for the smallness of the cosmological constant. He believed that the fundamental features of the physical world stem from grand, overarching principles that commend themselves by their beauty, economy, and explanatory power. He would have searched for a deep reason why the great cosmic tug-of-war between attraction and repulsion should be a close-run thing rather than an overwhelming victory for repulsion.

So the challenge remains. The term that Einstein considered, discarded, reconsidered, and then abandoned must be reconsidered again. Attempts to explain the magnitude of this term convincingly have so far defeated the world's finest theoretical physicists. It will probably require another Einstein to provide the answer.

Einstein in the Twilight Zone

LAWRENCE M. KRAUSS

> LAWRENCE M. KRAUSS, the Ambrose Swasey Professor of Physics and a professor of astronomy at Case Western Reserve University, bases his research chiefly on the relations between quantum phenomena at fundamental scales and cosmology. He is also the author of a number of popular science books, including the best-selling *The Physics of Star Trek, The Fifth Essence, Atom,* and most recently *Hiding in the Mirror.*

In the process of writing my most recent book, *Hiding in the Mirror,* about our continuing love affair with extra dimensions, I had the opportunity to view an old *Twilight Zone* episode in which a little girl disappears behind a wall into the "fourth dimension." The hero of the program was an intrepid physicist, who, with a bit of chalk and a swift and clear mind, saved the girl, her father, and her dog just before the portal she had fallen through closed up forever.

I realized after watching the program that it contained a repressed childhood memory: part of the reason I first decided I wanted to became a physicist was so that I could become such a scientist-hero. I also realized that there probably would never have been such a character on television at all were it not for Albert Einstein, who created the modern image of the scientist both in touch with the most esoteric mysteries of the universe and concerned about the human condition. Einstein was undoubtedly the twentieth century's scientist superhero.

Along with its spectacular impact on popular culture, Einstein's legacy is everywhere throughout the scientific community. Generations of ambitious young Jewish kids like me decided to ignore parental advice to go to medical school, forsaking financial security for the opportunity to become theoretical physicists like Einstein, with (as my mother disdainfully moaned when I told her of my plans) "chalk dust on their jackets." In spite of such impassioned admonishments, the allure of the idea of sitting alone in one's study in the middle of the night and being the first person in history to understand some vital aspect of creation was just too tantalizing. But Einstein's style did not influence only budding scientists, it dramatically affected working scientists as well: for example, if today you go to the Institute for Advanced Study, where Einstein worked during his last decade, you will find that the highest regard is paid to those who give talks using blackboard and chalk instead of computers armed with PowerPoint.

Einstein's imprint, of course, includes substance as well as style. For the last thirty years of his life, he worked—largely alone and ultimately fruitlessly, it appears—on a unified theory of all interactions. In this work he was hindered by the fact that when he began his research two of the four known forces in nature had not even been discovered, much less explained. In the intervening years, we have successfully built theoretical frameworks to describe and correctly predict all phenomena associated with the three nongravitational forces: electromagnetism, the weak nuclear force, and the strong nuclear force. Gravity, alas, has thus far not fully yielded to our efforts.

The stumbling block has also been particularly Einsteinian, in that it has involved Einstein's famous nemesis, quantum mechanics. What has thus far proved impenetrable is deriving a fully consistent quantum-mechanical formulation of general relativity that allows predictions that can be made and tested. Part of the problem is that gravity is simply so weak—compared with the other forces between fundamental particles at scales we can currently measure—that it is difficult to find opportunities to probe for quantum effects. The other part is that the very nature of general relativity seems to imply that standard methods by which the other forces in nature have been made compatible with quantum mechanics will produce nonsensical results when applied to gravity.

Two solutions to this problem have been proposed, both of which, again, reflect the Einsteinian legacy. The first has involved a generation-long effort to derive a new type of unified theory that might encompass gravity and the other forces in nature while at the same time producing a consistent quantum theory. This theory, of necessity, would require that general relativity be incorporated into a yet more general formulation that would overcome the standard obstacles to quantization—namely, the appearance of an infinite number of infinitely large terms in the predictions of the theory. One such candidate theory has been proposed, which over the years has been given a number of names, beginning with string theory and moving along to M-theory, an overarching version of string theory that features membranes, with the M standing for, as you like, "membrane," "mother," "matrix," or "mysterious."

Remarkably, if one assumes that, at some basic level, what we have previously considered to be point particles instead reflect the various states of excitation of vibrating strings, then several miraculous results arise: First, such theories require a particle with the properties of the graviton, the particle that, in a quantum-mechanical theory of gravity, would transmit the gravitational force. Second, with such theories it is possible—at least in principle—to avoid the infinities that otherwise plague quantum versions of general relativity. But this first solution comes at a cost: string theory and its successors are inconsistent in a mere four dimensions and require a total of ten, eleven, or twenty-six dimensions to resolve these inconsistencies.

What's worse is that such theories themselves are so complicated that we have yet to fully understand precisely what kind of effective four-dimensional world they might produce on the macro level—or indeed, whether they would produce any such world. What to do with all the extra dimensions is also, at present, completely unexplained. Are they hidden by being curled up into tiny balls, too small to detect via current experiments? Or is it possible that the very notion of "dimensionality" itself is an antiquated concept and inappropriate to apply to such a theory? Thus, while many attractive theoretical results have been derived, they remain far outside the empirically testable world of physics. We will have to wait either for some wonderful new ideas or some lucky new experiments to know if this approach is on the right track.

The second way to solve the problem is one that Einstein himself would have found more satisfying. He never fully accepted the stochastic nature of quantum-mechanical measurements, and he longed for a theoretical modification that would make not just the predictions of quantum mechanics but its physically observable measurements completely deterministic. Recently, several influential physicists, including the Nobel laureate Gerard 't Hooft, have been advocating a reexamination of the possibility that it is quantum mechanics, not gravity, that is at the root of the problems associated with general relativity. At the moment, the power, measured by the comparative numbers of physicists working on these different approaches, strongly resides in M-theory. But time will tell.

Einstein's legacy does not haunt merely the study of elementary-particle interactions; it also dominates the current study of the dynamics of the largest objects in the universe, including the entire observable universe itself. In 1998 the expansion of the universe, first measured by Edwin Hubble in 1929, was seen to be accelerating, not decelerating. This is preposterous, of course, because gravity is, for all normal types of matter and radiation, universally attractive; thus the mutual attraction of galaxies and clusters should slow any expansion. In 1917, in order to explain the incorrect but apparent fact that the universe is static on large scales, Einstein introduced a quantity into his equations that might counter the attractive force. This cosmological constant, as it became known, would result in a universal repulsive force throughout empty space that might counterbalance the gravitational attraction between distant bodies.

As it happened, Einstein made a mathematical error, and his cosmological constant could never have resulted in a static stable universe; however, the discovery that the universe is expanding obviated the need for such a repulsive force, since in an expanding universe gravity can be purely attractive, working to slow the expansion and perhaps ultimately reverse it. Einstein was reported to have said that introducing the cosmological constant was his "greatest blunder."

It now seems that his biggest blunder might have been abandoning this term too quickly. We currently have no idea what might be responsible for the observed acceleration of the universe, but the best bet is something like a cosmological constant. We now understand this term in a different way than Einstein did. It turns out that if one allows empty space to have energy, then this will automatically result in the appearance of a cosmological constant. And the laws of quantum mechanics, when combined with relativity, imply that such a term should be present—that is, that empty space should have energy. The only problem is that when we try to estimate how much energy empty space should have, we derive a number about 120 orders of magnitude larger than is allowed by observations. Clearly, there is something profound that we do not yet understand at the interface of quantum mechanics and gravity. Whether or not it will require a fundamental revision of our understanding of the theory that Einstein discovered, it is likely that the ideas he introduced will be at the heart of the matter.

For scientists like me, however, it is not merely the incredible breadth and depth of Einstein's contributions to physics that have elevated him to superhero status. Einstein was also vitally concerned with the human condition, and he wrote prolifically about social issues, elevating the role of citizen-scientist to a new level. This is all the more significant because by nature Einstein seemed to shy away from public displays. But he recognized that whether he liked it or not, there was a vital connection between the results of the scientific enterprise and our common human welfare. He was also intelligent enough to recognize the public power of his celebrity without taking it seriously in his private life. So it was, for example, that in the summer of 1939, when the Hungarian-born physicist Leo Szilard and others drafted a letter for him to send to President Roosevelt urging him to sponsor research into the development of what would eventually become nuclear weapons, Einstein agreed to do so, in spite of his long-standing pacifist tendencies. His direct experience of the Nazi regime had convinced him that the danger of not acting at that time was greater than the danger of acting.

As Einstein's fame increased over the years, he was called upon to lend his name to more and more causes, but he stuck firmly to his core beliefs in spite of pressures to do otherwise. One of the most profound of those beliefs was that it was his scientific achievements that later scholars should concentrate on and not his personal life. It is an interesting sociological fact that societies associate artists with their art in different ways; the ancient Greeks, for example, viewed the products of human creativity as separate from those who produced them. Thus, unlike in the present era, there was no obsessive cult of personality surrounding celebrity. I am aware that in writing now about Einstein's legacy I may appear to have fallen prey to our current generation's fixation on the famous. I want therefore to return to what Einstein himself recommended to posterity, which was that it was the ideas rather than the man that we should celebrate. And when we contemplate the legacy of the man whom modern society has stereotyped as the quintessential absentminded but brilliant scientist, we should realize that it was fully the power of his ideas that forever enmeshed his person in so many aspects of our popular culture and our scientific enterprise. Yes, it is true that Einstein didn't wear socks underneath his formal wear, and that he may have had numerous affairs during his lifetime, but his science is what truly produced his legacy. In the midst of celebrating his celebrity over the course of this hundredth anniversary of the development of special relativity, we should not lose track of the way he wanted to be remembered.

And yet . . . while in one sense I regret the cultification of Einstein the man, at the same time I can't help but think that in a society where heroes are more often than not chosen on the basis of their hand-eye coordination or their good looks or even their propensity for mayhem and violence, it is marvelously refreshing to have at least one scientist superhero, whose ideas have permeated society so profoundly that they have found a place even in *The Twilight Zone*.

No Beginning and No End

PAUL J. STEINHARDT

PAUL J. STEINHARDT is the Albert Einstein Professor in Science at Princeton University, with appointments in the Departments of Physics and Astrophysical Sciences. His research spans problems in particle physics, astrophysics, cosmology, and condensed-matter physics. He is one of the architects of the inflationary model of the universe, a modification of the standard Big Bang picture that explains the homogeneity and geometry of the universe and the origin of the fluctuations that seeded the formation of galaxies and large-scale structure. He pioneered the study of "quintessence," a dynamical form of dark energy that may account for the current cosmic acceleration. In 2002, Steinhardt received the P. A. M. Dirac Medal from the International Centre for Theoretical Physics.

Albert Einstein passed away when I was two years old, so for most of my life he has been a distant legendary figure. Since moving to Princeton in 1998, however, I have felt his presence all around me. A few miles from my office at Princeton University is the Institute for Advanced Study, of which Einstein was a founding member. I pass Einstein's white frame house on Mercer Street every day on the way to the university. A fireplace above which is inscribed Einstein's famous quote "*Raffiniert ist der Herr Gott aber boshaft ist er nicht*" ("God is subtle, but He is not malicious") is across the street in the building that used to serve as home to our Physics Department. A large bust of Einstein sits in the lounge where I get coffee each morning. As I return to my office, I pass a wall of photographs that show Einstein meeting with other renowned physicists in Princeton. On my bookshelf are hundreds of physics texts grounded in Einstein's foundational contributions. Even the title of my Princeton professorship bears Einstein's name. But what has been most striking is how my research has changed direction and my thoughts have been turning more and more to Einstein and his insights on the nature of the universe.

Perhaps that does not seem so strange for someone who studies the origin and evolution of the universe. After all, the cosmos is governed by gravity, so Einstein's general theory of relativity must underlie any contemporary model of the universe. My preoccupation with Einstein, though, has less to do with any technical contribution than with his basic instincts and philosophical outlook regarding the universe as a whole.

Einstein's views were revealed in his first attempt to apply general relativity to the cosmos. The year was 1917, shortly after he had introduced his revolutionary theory of gravity (but before Arthur Eddington's 1919 eclipse expedition had verified it). Einstein's uncanny intuition had been famously successful when applied to almost every domain of physics, so he confidently put forward a bold new model of the universe. The paper set a new standard for the field. Many technical aspects introduced there continue to be an integral part of cosmology.

Yet Einstein's record in cosmology is generally regarded as mixed. His work was based on his conviction that ours is a static, unchanging universe, yet in the next decade that possibility had been disproved. Following the work by Vesto Slipher and Edwin Hubble showing that the universe is expanding, most cosmologists discarded the static concept in favor of an alternative that ultimately developed into today's standard hot Big Bang model. (Fred Hoyle, Hermann Bondi, and Thomas Gold made a last-ditch attempt to save Einstein's picture of an eternal universe by hypothesizing a steady-state model, but this idea was defeated by the discovery of quasars and the cosmic background radiation in the 1960s.) The current consensus is that Einstein was wrong: the universe had a definite beginning 14 billion years ago and has a strange, uncertain future. Cosmologists excuse Einstein for his "blunders" on the grounds that he had virtually no astronomical observations on which to base his judgment and no knowledge about the forthcoming breakthroughs in fundamental physics that would bolster the hot Big Bang model.

But almost ninety years later, I find myself wondering whether Einstein's intuition was as far wrong as most cosmologists suppose. The recent discovery of cosmic acceleration and new ideas about space-time have led me to reconsider the words of the master.

The realization that the expansion of the universe is

accelerating has led to a rebirth of interest in the cosmological constant, which Einstein first introduced in his seminal 1917 paper as a device for making the universe static. The universe could not be static with matter alone, he realized, because of matter's gravitational self-attraction. So anxious was he to preserve his static universe that he decided to mar his beautiful equations of general relativity to combat the tendency of matter to collapse. He introduced a "cosmological constant" that produces a gravitationally repulsive counterforce, which can be finely tuned to keep the universe in balance. After Hubble and Slipher disproved the static model, Einstein withdrew his modification of general relativity, declaring it to be his greatest blunder. At the time, the cosmological constant appeared to be headed for the dustbin.

Then, in the 1990s, a multitude of cosmological measurements—including observations of the cosmic background radiation, the distribution of galaxies, and the light from distant supernovas—reached a common conclusion: the expansion of the universe is accelerating. The acceleration implies that most of the energy in the universe consists of a gravitationally self-repulsive ingredient similar to (or perhaps precisely equal to) the cosmological constant—an energy component dubbed dark energy. Einstein's greatest blunder is suddenly in vogue, and he is once again heralded as a visionary.

Yet in today's Big Bang model, the role of dark energy is diminished compared with what Einstein had in mind. It does not prevent the universe from having a beginning. It plays no role in the billions of years of evolution during which matter and radiation were created and the distribution of energy throughout space was established. It has no relation to the formation of the first galaxies, stars, and planets. It has no effect on the evolution of the universe until ten billion years of dramatic evolution have passed. Then, when dark energy finally becomes significant, it does not lead to a static universe, as Einstein had envisioned. Instead, for the foreseeable future, dark energy causes the expansion to accelerate, gradually turning the universe into a vacuous wasteland.

What irony that dark energy has been revived and Einstein's intuition on this count has been vindicated, but in a context so antithetical to his original dream! Or could it be that the discovery of dark energy has deeper significance? Could this be a sign that Einstein was closer to the truth in 1917 than we are today?

I find myself asking these questions because during the last few years Neil Turok, at Cambridge University, and I have been developing a new competitor for the hot Big Bang model. Our ambition at the outset was to construct a model as different as possible from the standard picture and whose predictions agree with current observations to the same exquisite precision. When we started down this path, we did not know where we were headed, and the chances of success seemed minuscule, given the wealth of new astronomical observations that had ruled out all previous competitors. Nevertheless, we persisted and found a surprisingly simple, logical alternative we call the cyclic model. As it turns out, without intending to, we found an alternative to the hot Big Bang picture that is just as effective in explaining the universe we observe and yet comes closer to embodying Einstein's vision.

The cyclic universe turns the conventional hot Big Bang story topsy-turvy. As Einstein would have it, space and time exist forever. The Big Bang is not the beginning of time but instead a bridge to a preexisting era. The universe undergoes an endless sequence of cycles in which it contracts in a Big Crunch and reemerges in an expanding Big Bang, with trillions of years of evolution in between. The temperature and density of the universe do not become infinite at any point in the cycle; indeed, they never exceed a finite bound (about a trillion trillion degrees). The key events that fix the large-scale structure of the universe—the uniform distribution of matter and radiation, the absence of substantial curvature and warps, and the seeds for forming galaxies—do not occur in a period of inflation in the first instants after the Bang, as in the conventional picture, but instead through a period of slow contraction that occurs before the Bang.

After the Bang, each cycle proceeds through periods of expansion dominated first by hot radiation and then by cold matter. During these ten billion years, the primordial abundance of elements, the first atoms, the cosmic background radiation, galaxies, stars, and planets are all created. Then a period of dark-energy domination begins that lasts a trillion years or more. Space undergoes a period of accelerated expansion that spreads out the matter, entropy, black holes, and any other debris produced in the preceding cycle extremely uniformly, and any curvature or warps in space are ironed out. The remnants become so thinly spread that the once-twinkling universe approaches a nearly perfect vacuum. As expansion continues, the concentration of dark energy decreases. Finally the acceleration stops and the universe contracts in a Big Crunch. A bounce from Big Crunch to Big Bang automatically replenishes the universe with new matter and radiation, and a new period of expansion and cooling ensues. Quantum effects cause the bounce to occur in some places before others, leading to the production of small dips and peaks in the distribution of matter and radiation. These nonuniformities are the seeds for forming galaxies and larger-scale structures.

The motivation for the cyclic model was based on new ideas about space and time that have emerged from superstring theory, the leading candidate for a unified theory of the fundamental forces in nature. Here is another connection to Einstein: he was a pioneer in the search for unification, and he was particularly interested in theories, like general relativity, that reinterpret forces in terms of geometrical effects. More than a half century later, theorists are hotly pursuing Einstein's dream. According to this approach, all particles and forces are due to vibrations, rotations, and reconnections of a single type of geometrical entity—a one-dimensional string—in a space with ten dimensions. The cyclic model builds on this idea by giving dark energy, the Big Crunch, and the Big Bang a geometrical interpretation. In a powerful version of superstring theory known as M-theory, our familiar universe of three dimensions is embedded in a space with extra dimensions. (Here, for simplicity, we consider only one extra dimension.) A microscopic distance away from our three-dimensional universe lies another three-dimensional universe similar to ours. The two universes can move with respect to each other and interact, but we cannot reach out across the gap or see the other world, because all the particles we are composed of—electrons, quarks, photons, and so on—are constrained to move in only the usual three spatial dimensions.

According to the cyclic model, dark energy is due to a springlike force between the two universes that has two effects: First, when the two universes are far apart, the energy stored in the "spring" has a gravitational effect that causes our three dimensions to stretch at an accelerating rate; this corresponds to the accelerated expansion phase observed today. Second, the spring also draws the universes together and closes the gap, ultimately causing the two worlds to collide and bounce apart. New matter and radiation are created from the heat of the collision, which causes the three-dimensional universes to expand. This corresponds to the transition from a Big Crunch to a Big Bang. Quantum uncertainty effects cause the collision to occur at slightly different times in different places, leading to small nonuniformities in the distribution of matter and radiation that eventually seed the formation of galaxies. Our universe emerges from the collision with all the properties needed to explain the observed properties of the universe.

Although the cyclic model and the standard Big Bang picture propose radically different histories of the universe, they are surprisingly difficult to distinguish. Their predictions of uniformity, flatness, and tiny density variations are absolutely identical. However, there is one key difference: the two models predict gravitational waves with very different distributions. Gravitational waves are ripples in space that propagate through the universe at the speed of light. In standard Big Bang and cyclic pictures, a broad spectrum of gravitational waves is generated at the same time as the seeds for galaxy formation. However, the variation of the amplitude with wavelength is radically different in the two pictures. A number of experiments on the ground, in airborne balloons, and aboard satellites will be conducted in the next decade to search for these primordial gravitational waves, and we will thus be able to decide which history of the universe is correct.

As I review the cyclic model, what strikes me most is how close it comes to Einstein's vision without contradicting any of the observations made in the last nine decades. Though I never met him, I feel we have developed a close intellectual bond. Most cosmologists have become so conditioned to believing that the Big Bang is the start of the universe that they discount the cyclic possibility. Einstein would be more empathetic. Long after he accepted the overwhelming evidence for cosmic expansion, Einstein emphasized, in the 1945 edition of *The Meaning of Relativity,* that "one may not conclude that 'the beginning of the expansion' must mean a singularity."

Einstein invented dark energy in the form of a cosmological constant to keep the universe unchanging over time. He would likely appreciate that dark energy can also play the central role in keeping the universe cycling. In some ways, this use of dark energy is more magical. At one stroke, it clears out the debris created in previous cycles, converts itself from an accelerating force to a contracting force, drives the universe toward a Big Crunch, and, for technical reasons that cannot be related here, ensures that the universe stably maintains regularly repeating cycles.

Einstein may have preferred a static universe, but a cyclic universe has a similar philosophical appeal. Both imply a universe with no beginning and no end. "Static" means that the average properties of the universe are the same from moment to moment. "Cyclic" can also be understood as maintaining the same average physical conditions, provided the average is taken over many bounces. In this sense, the cyclic model is the best compromise between Einstein's vision and observational reality. Perhaps by 2017, the hundredth anniversary of Einstein's seminal paper on cosmology, experiments will tell us whether nature has chosen this compromise.

Where Is Einstein?

MARIA SPIROPULU

MARIA SPIROPULU is an experimental physicist and a former Fermi Fellow at the University of Chicago's Enrico Fermi Institute. Born and educated in Greece, she developed an early interest in experimental physics, working as an undergraduate at BESSY (the Berlin Electron Storage Ring Company for Synchrotron Radiation) and at CERN. She moved to the United States in 1993 to pursue a Ph.D. at Harvard. At the Collider Detector at Fermilab, she worked on silicon sensors for the detection of high-energy particle decays and on supersymmetric searches, using the "blind" data method for the first time in analyzing hadron collider data. Spiropulu has given many public talks on physics, at venues ranging from the University of Chicago to the Wheeler Opera House in Aspen to radio shows and science documentaries.

Paris, May 28, 2004. In the Métro. There is a huge picture of Albert Einstein above the car's sliding doors. With his Einstein hair and eyes. Next to it, a picture of a young, sleek, twenty-first-century yuppie with a broad smile and businesslike demeanor, and underneath the legend (in French): "He is not available—but our [name of such and such a company's employee] is!"

Russin, early July 2004. A small restaurant outside Geneva, in the middle of the vineyards. A red-haired waitress is impatient to get rid of us. There has been a string workshop at CERN (the European Organization for Nuclear Research), following one in Paris and to be followed by a satellite meeting, once again at CERN. I'm with friends and colleagues—string theorists as well as the classic type of radicals in present-day physics—from the U.S. of A. I am new around here. Very new. But I took the advice of a Swiss colleague and drove at random through the countryside close by the lab, and sure enough I found a bunch of *Auberges d'Or* and *d'* this and that. We have spent the last three hours of the dinner talking about current affairs in physics and debating and pondering over Einstein. In particular about Einstein, David Hilbert, and the level of independence or interdependency in their coming up with the gravitational equations that would solve Mercury's orbit. This is neither a short nor a tepid conversation among physicists—especially theorists, as it turns out. We are the last to leave—virtually thrown out of the place.

Vienna, mid-July 2004. The Physics of LHC (Large Hadron Collider) workshop. Last talk, given by Chris Quigg, of Fermilab. The summary's last sentence (not an exact quote but as close to it as I can register): "Where will the next Einstein lead scientific thinking?"

CERN, end of July 2004. When I ask Savas Dimopoulos,

a particle theorist and model builder from Stanford, what inspired him to become a physicist, he recalls reading two biographies of Einstein as a kid in Athens.

That summer, I kept receiving e-mail with "Einstein" in the subject line. Mostly because it was almost the centenary of what is known as his miracle year of 1905. One of the most enticing was a tip that from August 8 to August 11, 2004, the Aspen Center for Physics would be holding (in the words of the *Aspen Times*) an "ambitious and high-profile public conference that will investigate the shock waves that Einstein and his work sent through the worlds of science, society, culture, intellectual affairs and beyond." A very high price tag ($700) went with the event. I asked a colleague who was planning to attend to let me know what it was all about. Was there something to learn about Einstein, his work, his life, his genius? What were the panels about, other than panelists discussing the nature of their own genius? But, as I suspected, it was the kind of theological cult ceremony in which people mingle and schmooze and the binding is their god—which, for this particular crowd, is Einstein. Glitz, glitterati, high intellectual performances. And the science journalist Dennis Overbye, who, as it turns out, knows more than anyone else I know about Einstein the man and Einstein the scientist.

I had chatted with Dennis the year before, when he was visiting Fermilab's Tevatron, about Einstein and about Dennis's new book, *Einstein in Love.* He told me about his seven-year adventure in Europe tracing and piecing together Einstein's life and works. A nontrivial task: legal affairs, official permissions to dig into archives, interpreters, and so on. I don't remember if I ever asked him exactly why he had done this—nor his answer if I did. (I found out later when I read the book.)

Einstein constituted somewhat of a family problem ever since the day I proclaimed myself a physicist. My father did not see in this research business much of immediate use to anybody, and in particular himself. Whenever I came home from college, he would ask the same question: "Have you done anything beyond Einstein, a century later? No? Not yet? Well, when will we see some progress in physics?" He still asks the same question.

I recently moved from Fermilab, outside Chicago, to Europe to work for CERN, a grand laboratory I knew as an undergraduate when I spent some time working for the experimental support section called Technical Assistance 2. And fell forever under the spell of particle physics. Geneva was described to me as the metropolis of Swiss/ Franco/German stuffiness, so six months before I arrived I started preparing myself for the transition. I thought, "If there is a lot of rubbish, well, Einstein went through this"—and the thought was comforting, even empowering. Inspired by Dennis's book, which I had just finished reading for the second time, I was excited by the opportunity to visit Bern and check the neighborhood around the patent office, and run around Zurich to the same cafés near the Polytechnic where Mileva and Einstein hung out in a circle of intellectuals, Einstein playing the violin at times.

I went to Bern all right, but I didn't have the time to look for the patent office—or anything else, really. I went to submit a research proposal to the Swiss National Science Foundation. Soon afterward, I learned that this was not the way things worked here. You could not just come in fresh and new, put your ideas on the table, and expect to get research grants. No! Without a history within the system and the appropriate network of people at the high research/management level, I found that my proposal was guaranteed to go directly into the trash bin, and I was given a friendly piece of advice: Don't try to change the system. I was clueless as to what "the system" meant, but now I'm getting hints and am rather close to at least characterizing it: It is no prettier than I understand it was when Einstein had to deal with it. In his case, he felt like an outsider because of his Jewish heritage. And this helped him to be a rebel in thinking about physics.

And rebel he was, particularly at the beginning of his career. A fearless and radical outsider with no mentors or guardian angels—or "champions," in today's physics language. A rather dramatic start for a clerk in a patent office, later proclaimed by the world as the launcher of modernity. I searched some of the physics publication databases and found some thirty papers authored or coauthored by Einstein. I searched the same archives for papers that had "Einstein" in their title and found 14,116. The physics subcategories under which the papers are filed include algebraic geometry, astrophysics, atomic physics, biological physics, chaotic dynamics, chemical physics, differential geometry, dynamical systems, exactly solvable and integrable systems, general relativity and quantum cosmology, geometric topology, high-energy physics: phenomenology, high-energy physics: theory, history of physics, mathematical physics, mesoscopic systems and the quantum hall effect, metric geometry, pattern formation and solutions, physics education, plasma physics, quantum algebra, quantum physics, soft condensed matter, space physics, statistical mechanics, and superconductivity. If you Google Einstein, you get more than 4.5 million references. The man has had a colossal cultural and societal effect. Fifty years after his death, every week there are hundreds, if not thousands, of news stories around the world touching on some aspect of his life and work. As a young physicist and researcher, I was getting tired of the fact that contemporary physics is largely synonymous with Einstein. And slightly disturbed that Einstein's "dream" of a unified field had become an advertising slogan used to get funding and promote physics in books, documentaries, interviews, and so on. But the truth of the matter is that the slogan is effective, and rightfully so, even today.

A recurrent question among physics researchers is, What would Einstein think of recent experimental results and theories in particle physics, gravitation, and cosmology? He was among the first to think seriously on all these issues, even at the beginning of the twentieth century. The accelerating universe/cosmological constant conundrum and the possibility of extra space dimensions were all topics that Einstein dealt with. Some think that the actual "Einstein's dream" was not a unified "theory of everything"—in particular, one describing gravity in a common framework with the rest of the physical phenomena—but rather a theory that would derive elementary particles and quantum mechanics from nonlinear solutions to classical field equations, and that soon after his success he decided to devote all his efforts to a problem he knew he couldn't (or wouldn't) solve.

And where is Einstein today? In our research, our studies, our experiments, our calculations. In college textbooks from physics to the history of culture. In our witticisms. In the hearts of our friends and colleagues who call in the middle of the night with some preposterous idea about how to solve a preposterous problem. In the hearts of the artists and businessmen and electricians and philosophers we meet who want to know what $E = mc^2$ really means as soon as they learn we are physicists. In intellectuals who know how to be cute and modest and play the media. In those who take sides. In every demonstration against nuclear weapons. In all young people who exhibit a slight disrespect for authority and all old people in authority who are radicals. In those who have fled from tyrannical and oppressive families and narrow-minded educators. In all our students who can solve a problem we cannot. In all the people we think of as innocent geniuses who take themselves not very seriously. In all stubborn and difficult physicists who think of physics problems as a reason for being. And in all those who tackle problems they know they can't or won't solve.

Acknowledgments

I wish to thank my publisher, Marty Asher of Pantheon Books, for his encouragement.

I am also indebted to my agent, Max Brockman, who recognized the potential of this book, and to Sara Lippincott for her thoughtful and meticulous editing.

ABOUT THE EDITOR

John Brockman has edited nearly twenty books and is the author of *By the Late John Brockman, The Third Culture,* and *Digerati: Encounters with the Cyber Elite.* He is the founder and CEO of Brockman Inc., an international literary and software agency; president of Edge Foundation, Inc.; and publisher and editor of *Edge,* a Web site presenting the third culture in action. A well-known computer and Internet entrepreneur and visionary, he and his work are frequently featured in the media.

A NOTE ON THE TYPE

This book was set in Minion, a typeface produced by the Adobe Corporation specifically for the Macintosh personal computer and released in 1990. Designed by Robert Slimbach, Minion combines the classic characteristics of old style faces with the full complement of weights required for modern typesetting.

Composed by
Stratford Publishing Services,
Brattleboro, Vermont
Printed and bound by
R. R. Donnelley & Sons,
Harrisonburg, Virginia
Designed by Virginia Tan